Aurélien Godfrin

AC Losses in High-Temperature Superconductor Tapes and Cables for Power Applications

Karlsruher Schriftenreihe zur Supraleitung Band 033

HERAUSGEBER

Prof. Dr. Tabea Arndt
Prof. Dr. rer. nat. Bernhard Holzapfel
Prof. Dr. rer. nat. Sebastian Kempf
Prof. Dr.-Ing. Mathias Noe

Eine Übersicht aller bisher in dieser Schriftenreihe erschienenen Bände finden Sie am Ende des Buches.

AC Losses in High-Temperature Superconductor Tapes and Cables for Power Applications

by
Aurélien Godfrin

KIT Scientific Publishing

Karlsruher Institut für Technologie
Institut für Technische Physik

AC Losses in High-Temperature Superconductor Tapes
and Cables for Power Applications

Zur Erlangung des akademischen Grades eines Doktors der Ingenieurwissenschaften von der KIT-Fakultät für Elektrotechnik und Informationstechnik des Karlsruher Instituts für Technologie (KIT) genehmigte Dissertation

von Aurélien Godfrin, M.Sc.

Tag der mündlichen Prüfung: 21. Juli 2020
Hauptreferent: Apl. Prof. Dr. Francesco Grilli
Korreferenten: Prof. Dr.-Ing. Mathias Noe, Prof. Bruno Douine

Impressum

KIT Scientific Publishing

Karlsruher Institut für Technologie (KIT)
KIT Scientific Publishing
Straße am Forum 2
D-76131 Karlsruhe

KIT Scientific Publishing is a registered trademark
of Karlsruhe Institute of Technology.
Reprint using the book cover is not allowed.

www.ksp.kit.edu

Print on Demand 2022 – Gedruckt auf FSC-zertifiziertem Papier

ISSN 1869-1765
ISBN 978-3-7315-1096-3
DOI 10.5445/KSP/1000131627

Acknowledgements

First of all, I would like to express my sincere gratitude to my supervisor, Apl. Prof. Dr. Francesco Grilli, for his continuous support, his patience, his motivation, his enthusiasm and his immense knowledge. Besides, his meticulous review of each of the chapters allowed me to clarify my writing. I could not hoped having a better supervisor for my PhD.

I would further like to thank the whole Institute for Technical Physics (ITEP) for the welcome and the excellent working conditions that were offered to me and especially, the managing director of the institute, Prof. Dr.-Ing. Mathias Noe for his advices and for being one of my co-examiners. These years spent at ITEP allowed me to learn a lot, to improve my skills, to work with brilliant scientists, to participate in conferences and workshops, and to encounter great people.

In addition, I would like to acknowledge my PhD fellows Dr. Sebastian Hellmann, Florian Erb, Dr. Carolin Heidt, Dr. Pattabhi Vishnuvardhan Gade, Santiago Ochoa, Matthieu Scannapiego and Cristian Gleason-González for the moments of joy and lough as well as the great time spent at ITEP and at Karlsruhe. I would also like to thank my colleagues Dr. Anna Kario, Dr. Roland Gyuráki, Dr. Eduard Demenčik, Dr. Victor Manuel Rodriguez-Zermeño for the scientific discussions and the interpretation of results. I am grateful to my labmates Andrea Kling, Andrej Kudymow, Rainer Gehring and Dr. Michal Vojenčiak (Slovak Academy of Sciences) for their help regarding the tuning of the experimental devices and the measurements.

I would like to express my acknowledgments to my former professors of the GREEN Laboratory, without whom these marvelous years of thesis would not have started. A special thank to Prof. Bruno Douine for his advices and for being one of my co-examiners.

Lastly, and most of all, I would like to thank my family and particularly my beloved wife Lucile, for her constant support and her love throughout my PhD despite the distance.

—— To my children, Gaspard and Célestine ——

Zusammenfassung

Aufgrund ihrer hohen Stromdichte, ihres guten Feldverhaltens und ihrer mechanischen Festigkeit werden Hochtemperatursupraleitern (HTS) in Form von beschichteten Bandleitern auf Seltenerdbasis für Großanwendungen wie Transformatoren, Magnete und Elektromotoren vorgeschlagen. Diese beschichteten Leiter weisen jedoch aufgrund ihres hohen Aspektverhältnisses mit einer Breite von mehreren Millimetern und einer Dicke in der Größenordnung von 1 µm große Wechselstrommagnetisierungsverluste auf, was ihre Verwendung in diesen Leistungsanwendungen behindert.

Ein üblicher Weg zur Verringerung der Wechselstromverluste von beschichteten Leitern besteht darin, den Supraleiter in schmale Filamente zu unterteilen. Die Laserablation ist eine der am häufigsten verwendeten Methoden zur Herstellung von Filamenten, jedoch wird dieser Prozess durch die in Leistungsanwendungen benötigte ausreichend dicke Kupferstabilisierung der Supraleiter erschwert. Der beste Weg um Filamente in kupferstabilisierten beschichteten Leitern herzustellen, wird noch entwickelt.

Der Schwerpunkt der Arbeit liegt auf zwei Themen. Das erste ist die Untersuchung der Herstellung von Filamenten auf kupferstabilisierten beschichteten Leitern, wobei die Filamente nach oder vor der galvanischen Beschichtung des Bandes erzeugt werden. In der gesamten Arbeit werden die mit diesen beiden Ansätzen hergestellten Proben als "Striated After Electroplating" (SAE) und "Striated Before Electroplating" (SBE) bezeichnet. Untersucht werden Proben mit unterschiedlicher Filamentanzahl (10 bis 60) und Kupferdicke (5 µm und 10 µm). Das zweite Thema ist die Anwendbarkeit der Filamentisierung zur Reduzierung der Wechselstromverluste von Kabeln, insbesondere der CORC®- und RACC-Kabel, die mit filamentisierten HTS-Bändern hergestellt werden.

Die Wechselstromverluste werden für Bänder mit der kalibrierungsfreien Methode gemessen und für Kabel mit der kalorimetrischen Methode, die für die Untersuchung von HTS-Bändern und -Kabeln geeignet sind. Beide Methoden arbeiten bei 77 K in einem weiten Frequenzbereich des externen Magnetfeldes, wodurch der Beginn der Kopplungsströme und die Bedeutung der Kopplungsverluste bestimmt werden können. Ergänzende Gleichstrommessungen des Querwiderstands zwischen Filamenten helfen, den Pfad der Kopplungsströme zu verstehen.

Dank der Vielzahl der getesteten Proben und Betriebsbedingungen bietet die vorliegende Arbeit einen direkten Vergleich der beiden Filamentprozesse in handelsüblichen HTS Bandleitern. Bei Bändern hilft die Filamentisierung, die Wechselstromverluste zu verringern, und je höher die Anzahl der Filamente desto geringer sind die Wechselstromverluste. Die Verlustreduzierung hängt jedoch stark vom Prozess des Filamentierens und von der Dicke des Kupfers ab. Bei den Wechselstromverlustsmessungen mit höheren Frequenzen wurde angenommen, dass ein Pfad vorhanden ist, auf dem die Kopplungsströme leicht fließen können. Dies wurde durch SEM-Bilder und durch Querwiderstandsmessungen bestätigt. Bei beiden Filamentprozessen (SAE und SBE) ermöglicht eine dünnere Kupferschicht eine effektivere Verringerung der Wechselstromverluste und eine Begrenzung der Kopplung. Die Verwendung von SAE-Bändern in Anwendungen kann durch Delaminierungsprobleme erschwert werden. Weitere Untersuchungen zu diesem potenziellen Problem sind daher erforderlich. Wenn man jedoch die Komplexität und Dauer der beiden Prozesse und die Ergebnisse der Wechselstromverluste berücksichtigt, scheint der SAE-Prozess die bevorzugte Option zu sein.

Die Verwendung von Filamenten reduziert auch die Verluste von CORC®-Kabeln. Die Verlustreduzierung von CORC®-Kabeln war jedoch geringer als die für einzelne filamentisierte Bänder. Dies ist wahrscheinlich darauf zurückzuführen, dass die inneren Schichten durch die äußeren Schichten abgeschirmt sind. Die Wechselstromverluste wurden bei dem aus filamentisierten Teilleitern hergestellten RACC-Kabel leicht reduziert. Diese Tatsache ist auf die Filamentgeometrie des RACC-Kabels zurückzuführen. Das Filament, das sich an einer der Kanten des Teilleiters befindet, geht bei der Hälfte der Transpositionslänge nicht zur gegenüberliegenden Kabelkante. Die Wechselstromverluste von CORC®- und RACC-Kabeln mit ähnlichem kritischen Strom zeigen, dass die CORC®-Kabel für einen bestimmten angelegten Strom geringere Wechselstromverluste aufweisen als die RACC-Kabel. Dies erklärt sich aus der Tatsache, dass die Teilleiter des RACC-Kabels einem senkrechten Magnetfeld ausgesetzt sind, während die des CORC®-Kabels einem Feld mit verschiedenen Ausrichtungen ausgesetzt sind. Wenn die Verluste jedoch durch die Stromdichte des Kabels geteilt werden, weisen die RACC-Kabel geringere Wechselstromverluste auf. Dies erklärt sich aus der Tatsache, dass der formgebende Kern, der in CORC®-Kabeln zum Aufwickeln der Bänder verwendet wird, die technische Stromdichte verringert. Wenn der Durchmesser dieses Kerns verringert werden kann, kann das CORC®-Kabel wettbewerbsfähig werden.

Abstract

Large scale applications such as transformers, magnets, and electric motors are being proposed with high-temperature superconductors (HTS) in the form of rare earth-based coated conductors thanks to their high current density, good in-field behavior and mechanical strength. However, these coated conductors have large AC magnetization losses because of their high aspect ratio between the width (typically 4-12 mm) and the thickness (order of 1 µm), which hinders their use in those power applications.

A common way of reducing the AC losses of coated conductors is by striating the superconductor into narrow filaments. Laser ablation is one of the most used methods for making filaments. However, superconductors in power applications need a sufficiently thick copper stabilization and the best way of introducing filaments in copper-stabilized coated conductors is still to be found.

This thesis focuses on two topics. The first is the investigation of producing filaments on copper-stabilized coated conductors, with striations made after or before electroplating the tape. Throughout the thesis, the samples produced with these two approaches are called Striated After Electroplating (SAE) and Striated Before Electroplating (SBE). Samples with different numbers of filaments (from 10 to 60) and copper thicknesses (5 µm and 10 µm) are investigated. The second topic is the applicability of the striations for reducing the AC losses of cables, in particular the CORC® and RACC cables, which are made with HTS striated tapes.

The AC losses are measured with the calibration-free method for tapes and calorimetric method for cables, which are adapted for the investigation of HTS tapes and cables. Both methods operate at 77 K in a wide range of frequencies of the external magnetic field, which allows determining the onset of coupling currents and the importance of the coupling losses. Complementary DC measurements of the transverse resistivity between filaments help to understand the path of the coupling currents.

Thanks to the variety of tested samples and operating conditions, the present work provides a direct comparison of the two striation processes in commercially available HTS coated conductors. For tapes, making filaments helps reduce the AC losses and the AC loss reduction is higher with higher number of filaments. However, the loss reduction strongly depends on the striation process and on the copper thickness. The presence of a path where the coupling currents can easily flow was supposed, when higher AC losses were measured with higher frequencies, and was confirmed by SEM images and by transverse resistance measurements. For both striation processes (SAE and SBE), a thinner copper layer allows reducing more effectively the AC losses and limiting coupling. The use of SAE tape in applications can be hindered by delamination problems. Further investigations on this potential problem are therefore necessary. But if one considers the complexity and duration of the two processes and the AC loss results, the SAE process seems to be the preferable option.

Using filament reduces the losses of CORC® cables as well. However, the loss reduction of CORC® cables was smaller than that measured for single striated tapes. This can probably be ascribed to the fact that the inner layers are shielded by the outer layers. The AC losses were only slightly reduced with the RACC cable made by striating strands. This is due to the filament geometry on the RACC cable. The filament that is located on one of the edges of the strand does not go, at the half of the transposition length, to the opposite cable edge. The AC losses of CORC® and RACC cables with similar critical current show that, for a given applied current, the CORC® cables have smaller AC losses than the RACC cables. This is explained by the fact that the strands of RACC cable are exposed to a perpendicular magnetic field whereas those of the CORC® cable are exposed to a field of various orientations. However, if the losses are normalized by the cable's current density, the RACC cables have smaller AC losses. This is due to the fact that the former used in CORC® cable to wind the tapes reduces the engineering current density. In addition, if the diameter of the former is reduced, then the CORC® cable can become competitive.

Table of Contents

1 Introduction, Motivation and Outline of Work

Rare earth-based coated conductors are currently the most promising high-temperature superconducting (HTS) tapes by virtue of their high current density, in-field behavior and mechanical strength. Numerous large scale applications, which include power cables [MSB⁺07], fault current limiters [NS07, HAHN17], and magnets [HKK⁺19] for different uses [HHJ⁺04], are being proposed, with HTS coated conductors as basic current-carrying element.

One of the problems which still hinders the widespread use of HTS coated conductors is their relatively high AC losses, which constitutes a serious refrigeration burden [ARZH⁺11, Ain19]. This dissipation occurs in superconductors not only when they work in purely AC conditions (like in AC power cables or in transformers), but every time there is a time-variation of the magnetic field. As a consequence, even superconducting magnets operating in DC might be affected by the "AC loss problem" when they undergo current ramps for charging and discharging. AC ripples, i.e. small oscillations of otherwise constant currents or fields, are also a source of dissipation.

HTS coated conductor are particularly prone to developing high AC losses because of the very high ratio between the width (in the order of several mm) and the thickness (in the order of 1 μm) of the superconducting layer. This very high aspect ratio makes the magnetization losses – i.e. the losses caused by an external magnetic field – particularly high when the field is oriented perpendicular to the flat face of the tapes. Here is an example to understand the involved numbers and orders of magnitude. A 12 mm wide coated conductor, with critical current of 350 A (in self-field at 77 K, similar to the coated conductors considered in this thesis), subjected to a perpendicular magnetic field of 20 mT (which is a very modest amplitude) at 50 Hz generates a power dissipation of about $1\,\mathrm{W\,m^{-1}}$. One has to keep in mind that real applications (like transformers) require long lengths of superconductor tape and that this dissipation occurs at cryogenic temperature, which gives very high penalty factors. On top of this, one has to keep in mind that the efficiency with which such heat is removed from a cryogenic temperature (e.g. with the use of cryocoolers) is far from ideal, which further amplifies the penalty factor by which the AC loss value must be multiplied. This refrigeration burden can become so large as to completely cancel all the benefits coming from using superconducting technology.

In the particular case of coated conductors, we have already mention that the high AC losses are principally caused by the presence of a time-varying magnetic field perpendicular to the flat face of the tapes. In some cases it is possible to design the applications in a way that such field component is cancelled or at least minimized (e.g. with the use of flux diverters). However, this is not always possible in practice, and one has to find solutions to reduce the losses modifying the geometry or configuration of the HTS conductor.

A method for reducing the AC losses of coated conductor is by dividing the width of the tape into filaments. For a flat and very thin superconductor, the magnetization losses in perpendicular field scale with the square of the superconductor. So, by making N filaments, the losses can be reduced by a factor N [GK16]. This is only true, however, if the filaments are electrically uncoupled, i.e. if the cuts produced by the striation process are sufficiently "clean" so that the current cannot flow from one filament to the other. Unfortunately, this is difficult to realize in practice, especially if the HTS coated conductor needs to have an additional layer of copper on top for stabilization purposes.

The first main goal of this thesis is the investigation of different strategies for obtaining uncoupled filaments in HTS coated conductor with copper stabilization. In particular, it aims at determining whether it is better to striate the tapes before or after depositing the copper layer. The influence of the thickness of the copper stabilization is investigated as well.

The second goal of this thesis is to assess the applicability of the striation process to larger cables made of HTS coated conductor: in particular the CORC® and the RACC cables. These two cable types are characterized by having transposed strands, which is a necessity in order to avoid strand coupling at the terminals where the current is injected. CORC® and RACC cables made of striated coated conductors are investigated in terms of AC losses caused by the simultaneous presence of an AC transport current and an AC external magnetic field.

The thesis is organized in different chapters, as follows.

Chapter 2 recalls some fundamentals of superconductivity and discusses the different types of AC losses. Chapter 3 shows the preparation of HTS tapes and cables that were measured for this work and also how the AC losses were measured with respect to the type of sample (tape or cable). The maximum current that a superconductor can carry is one of parameters of superconductor that is important to know. This current called critical current depends strongly on the magnitude and the orientation of the magnetic field. Chapter 4 describes the experimental method to measure it as well as the measurement results of several HTS tapes and cables. Chapter 5 shows the investigation of the applicability of the laser ablation technique for producing copper-stabilized filamentized tapes. Also in this chapter the reduction of AC magnetization losses in coated conductors where the filaments are produced before and after the deposition of the copper is investigated. On the other hand the applicability of striations as means to reduce AC losses in HTS cables is investigated in chapter 6. Finally, conclusion and results of this work are presented in chapter 7.

2 Fundamentals and AC Losses in Superconductors

This chapter recalls some fundamentals of superconductivity and introduces the AC losses in HTS tapes and cables.[1]

2.1 Fundamentals of Superconductivity

2.1.1 Zero Resistance and Meissner Effect

The phenomenon of superconductivity was discovered by the Dutch physicist Kamerlingh Onnes in 1911 after he had managed to liquify helium three years earlier. While measuring the electrical resistivity of mercury as a function of temperature, he observed that the electrical resistivity of mercury suddenly dropped to an unmeasurable value at around 4.2 K, see Fig. 2.1.

Metals such as aluminum, lead, mercury, tin, titanium become superconducting, when they are cooled down below a specific temperature. The temperature is therefore strongly linked to the superconducting phenomenon. This specific temperature which defines the transition of the superconducting state to the normal state of the material is called the critical temperature T_c. The critical temperatures of some superconductors are listed in table 2.1. One can note that copper, silver, and gold are excellent electrical conductors at room temperatures, but they do not exhibit superconductivity, whereas some insulating material do.

Due to the absence of resistance below the critical temperature, a current induced in a superconducting loop is able to flow indefinitely in the material. These persistent currents have been observed to flow without any observable decay several years, which proves that the resistance is either zero or extremely small (lower than 10^{-26} Ω m). These persistent currents are employed in certain applications, such as MRI magnets.

[1] For this chapter, text from the following sources was used: [Rey15], [Ope], [SMM05].

Table 2.1: Values of the critical temperature T_c and critical magnetic field $B_c(0)$ at 0 K for various type-I superconductors. Source: [SMM05].

Superconductor	T_c (K)	$B_c(0)$ (T)
Aluminum (Al)	1.20	0.0105
Gallium (Ga)	1.08	0.0058
Mercury (Hg)	4.15	0.0411
Indium (In)	3.41	0.0281
Niobium (Nb)	9.26	0.1991
Lead (Pb)	7.19	0.0803
Tin (Sn)	3.72	0.0305
Tantalum (Ta)	4.47	0.0829
Titanium (Ti)	0.39	0.0100
Tungsten (W)	0.02	0.0001
Zinc (Zn)	0.85	0.0054

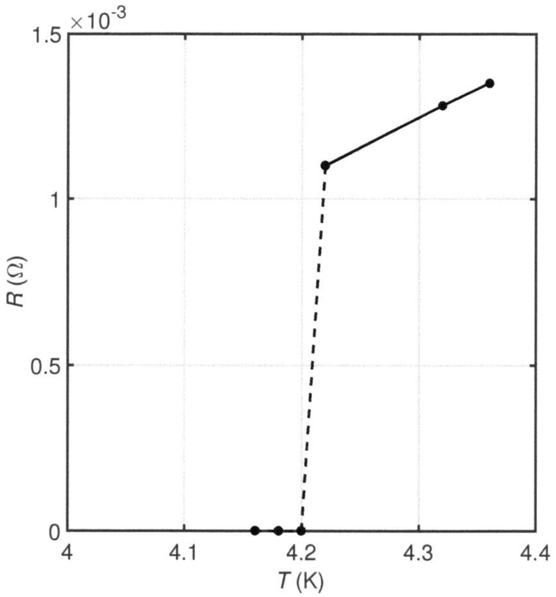

Figure 2.1: Discovery of the superconductivity (H. Kermerlingh-Onnes, 1911: resistance of mercury versus temperature).

Another important characteristic of superconductors was discovered by Meissner and Ochsenfeld in 1933. They discovered that the magnetic field is expelled from inside the superconductor. Thus the magnetic field inside the superconductor is canceled by screening currents appearing on the surface of the superconductor. The same state of magnetic field equal to zero is achieved whether the field has been applied before or after the material has been cooled below its critical temperature. Figure 2.2 shows this phenomenon – called Meissner effect – for a sphere.

The Meissner effect exists only up to a maximum critical field B_c, which decreases with increasing temperature, approximately as follows:

$$B_c(T) = B_c(0)\left[1 - \left(\frac{T}{T_c}\right)^2\right] \tag{2.1}$$

where $B_c(0)$ is the value of the critical field at zero temperature. It is thus possible to change the state of the sample from the superconducting state into the normal state not only by increasing its temperature, but also by subjecting it to a magnetic field higher than the critical field B_c.

Values of the critical field for various metals are referred in Table 2.1. One can note that the critical field values are quite low and can be easily reached in simple coils. This is why the first attempts of building superconducting coils failed.

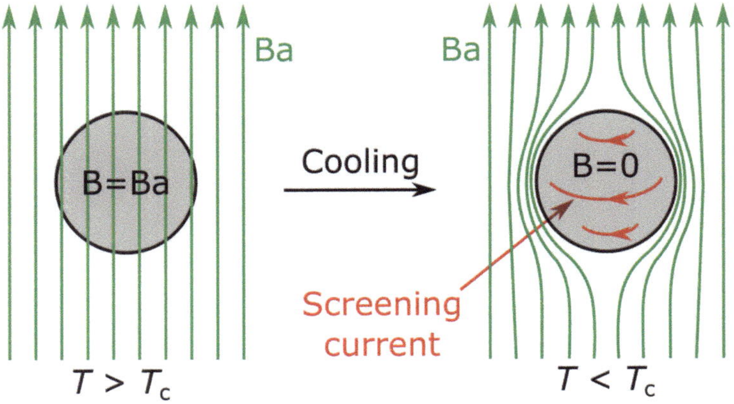

Figure 2.2: Meissner effect: expulsion of the magnetic flux inside a superconductor, due to the onset of screening current.

The existence of a critical magnetic field B_c indicates that there is also a maximum current the superconductor can carry. For example, the currents in a round aluminum wire of radius 1 mm cooled at 0 K ($B_c = 0.0105\,\text{T}$) that make the superconductor loses its superconducting state are larger than $I_c = 2\pi r B_c/\mu_0 = 52.5\,\text{A}$. So this maximum current depending of the applied field and the temperature is called the critical current I_c. This definition of critical current is known as Silbsee's rule and works well for pure elemental superconductors. However, as it will be discussed later, practical superconducting wires are made of alloys and compounds. Their critical current is determined by the presence of various types of crystal lattice defects and inhomogeneities in the material, which pin the magnetic flux lines thus preventing them to move and cause dissipation, and is therefore different from the simple estimation of Silbsee's rule.

In summary, a superconductor has three critical parameters: the critical temperature T_c, the critical magnetic field B_c or the critical current I_c. These define a critical surface as shown in fig. 2.3, which defines the limits within which the material is in the superconducting state.

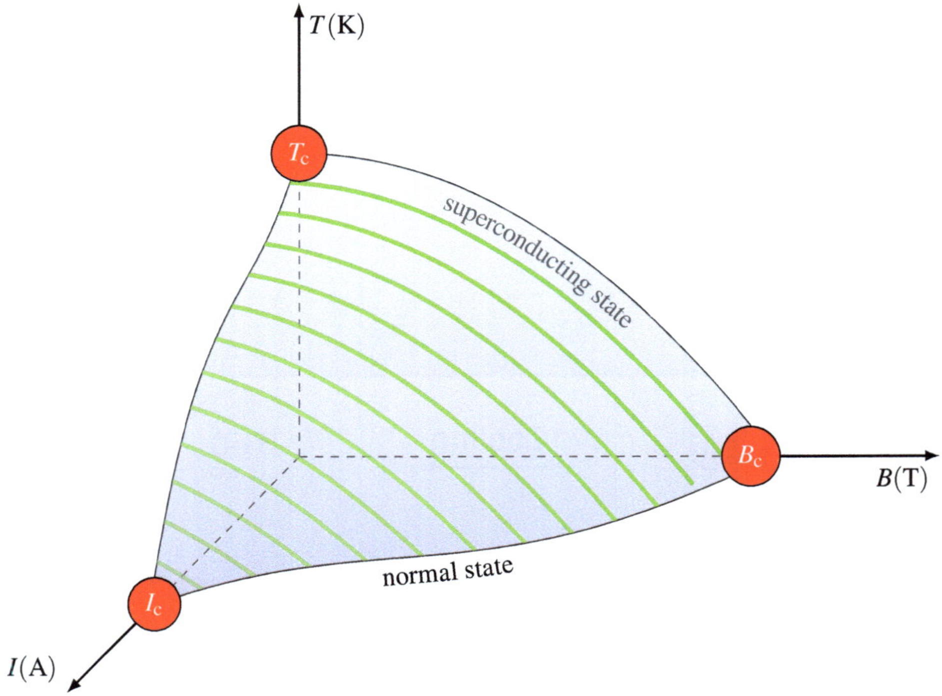

Figure 2.3: The critical surface given by the critical temperature T_c, the critical magnetic field B_c and the critical current I_c forms the limit between the superconducting and normal phases.

2.1.2 Type-I and Type-II Superconductors

For many years it was thought that the magnetic behavior described above was applicable to all superconductors. In 1957, Alexei Alexeyevich Abrikosov published a theoretical paper regarding another class of superconductors with different properties [Abr57]. These were later named type-II superconductors, to distinguish them from the conventional ones, which were then named type-I superconductors

Type-II superconductors are characterized by two critical magnetic fields, denoted as B_{c1} and B_{c2} and shown in Fig 2.4. When the external magnetic field is less than the lower critical field B_{c1}, the material is entirely superconducting and there is no flux penetration, as with type-I superconductors. For fields between B_{c1} and B_{c2}, the material is in a mixed state, also called vortex state. This name is given because, in this state, super-current vortices are created. In the mixed state, the magnetic field penetrates partially into the material in the form of vortices as shown in Fig. 2.5. These vortices are small tubular regions, each containing one quantum of flux:

$$\Phi_0 = \frac{h}{2e} = 2.068 \times 10^{-15}\mathrm{Tm}^2 \tag{2.2}$$

where e is the electronic charge and h the Planck constant. When the external field exceeds the upper critical field B_{c2} the flux penetrates completely and the superconducting state is destroyed, just as for type-I superconductors.

Table 2.2 gives some values of the critical temperature and the critical magnetic field B_{c2} at 0 K for several type-II superconductors. The values of B_{c2} are very large in comparison with

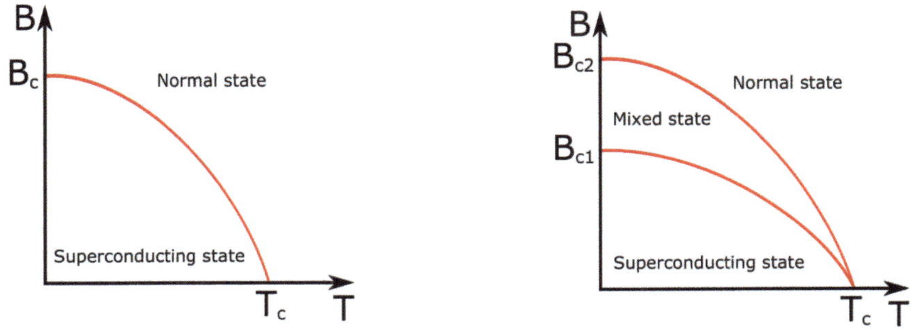

(a) Type-I superconductors.

(b) Type-II superconductors.

Figure 2.4: Critical magnetic field as a function of temperature for: (a) type-I superconductors, (b) type-II superconductors.

Table 2.2: Values of the critical temperature T_c and critical magnetic field B_{c2} at 0 K for various type-II superconductors. Source: [SMM05].

Superconductor	T_c (K)	$B_c(0)$ (T)
Nb_3Al	18.7	32.4
Nb_3Sn	18.0	24.5
Nb_3Ge	23.0	38.0
NbTi	9.3	15.0
PbMoS	14.4	60.0

those of B_c for type-I superconductors. For this reason, type-II superconductors are well suited for the construction of high-field superconducting magnets and other applications.

At the boundary between the normal and the superconductor state, the number density of superconducting electrons does not reach instantly its finite value, but it goes softly over a characteristic length ξ, the coherence length. This latter is an important parameter in order to differentiate type-I and type-II superconductors. The criterion for distinguishing type-I and type-II superconductors is given by the ratio known as Ginzburg-Landau parameter, $\kappa = \lambda/\xi$. A superconductor is type-I for $\kappa < 1/\sqrt{2}$ and type-II for $\kappa > 1/\sqrt{2}$.

As mentioned in [SMM05] "the coherence length and penetration depth both depend on the mean free path of the electrons in the normal state. The mean free path of a metal can be reduced by the addition of impurities to the metal, which causes the penetration depth to

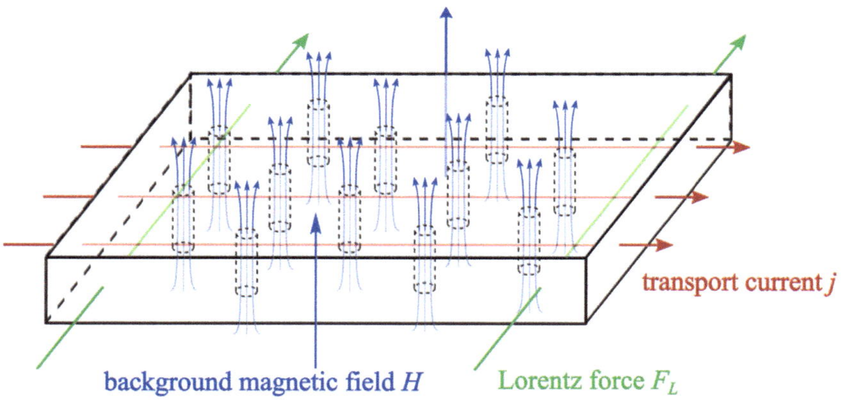

Figure 2.5: Schematic illustration of the mixed state, with the magnetic flux penetrating in the form of quantum vortices. When a transport current is additionally applied, the votices experience a Lorentz force that tends to move them. Source for the figure: [Kru14].

increase while coherence length decreases. Thus one can cause a metal to change from type-I to type-II by introducing an alloying element".

Type-I superconductors consist mainly of metals and non-metals that show some conductivity at room temperature, such as lead (Pb), mercury (Hg), tin (Sn), indium (In), and aluminum (Al). Type-II superconductors consist of metallic compounds and alloys, except for the elements vanadium, technetium and niobium. These type-II superconductors include some LTS, the high temperature superconductors (HTS) and the magnesium diboride (M_gB_2). The border between the two groups is 30 K, which is the limit predicted by the classic microscopic theory of superconductivity developed by Bardeen-Cooper-Schrieffer. This manuscript focuses on HTS materials, in particular REBCO tapes.

2.1.3 Flux Pinning and Macroscopic Currents

In type-II superconductors in the mixed state there is an interaction between the current and the 'tubes' of magnetic flux that thread through the normal cores.[2] The electrons will experience a Lorentz force, perpendicular to both the current density and the magnetic field. One can regard this as a mutual interaction between the electrons and the flux in the normal cores, as a result of which each normal core experiences a force that tends to move them, as shown in Fig. 2.5, and that causes energy dissipation.

This is avoided thanks to the presence of imperfections in the superconductor material, which effectively pin the normal cores in position - they provide a potential barrier to motion of the cores, so that the force on the cores must exceed a certain value before the cores can move. These pinning centers can also be artificially added. The more of these flux pinning centers that are present, and the greater the potential barrier they provide, the greater will be the current required to set them in motion, i.e. the greater the critical current.

The macroscopic behavior of type-II superconductors with strong pinning forces (also known as "hard" superconductors) is well described by the Bean model, proposed by C. P. Bean in 1962 [Bea62].

The basic principles can be understood by considering an infinitely long bar of superconducting material, subjected to a magnetic field applied along its longitudinal direction.[3] A transversal cross-section is represented in Fig. 2.6. The magnetic flux penetrates in the form of quantized vortices. This penetration creates a (spatial) gradient of magnetic flux: consequently, by Ampère's law $\nabla \times B = \mu_0 J$, a current must flow perpendicular to the flux lines. The local currents do not cancel out and a net local current flows inside the material. In the

[2]This part is taken from https://www.open.edu/openlearn/science-maths-technology/engineering-technology/superconductivity/content-section-4.3

[3]This is nicely described in [Rey15] and here we report the same example.

particular case shown in Fig. 2.6, the currents form a loop (blue arrows), which creates a magnetic moment opposing the applied field. As long as the average slope of the flux front $\mathrm{d}B/\mathrm{d}x$ exceeds $\mu_0 J_\mathrm{c}$, the Lorentz force pushes the flux lines out of the pinning "hole": the flux front advances, and the flux gradient correspondingly decreases. The process continues and stops only when $\mathrm{d}B/\mathrm{d}x = \mu_0 J_\mathrm{c}$, i.e. when the potential energy associated with the pinning centers equals the Lorentz force. If the amplitude of the applied field is increased, more flux lines enter the material and the front advances and stabilizes a little bit further inside the superconductor, but the slope of the front is still equal to $\mu_0 J_\mathrm{c}$. Also a gradient in the density of vortices is reluctant to any rearrangement. Thus, once a DC current is established in a hard superconductor, it will persist also after switching the driving voltage off (resistance-less circulation of an electrical current). However, in AC regime the vortices must move to follow the change of the magnetic field: the pinning force represents an obstacle and superseding it is an irreversible process: ramping an external magnetic field up and reducing it to zero leaves some magnetic flux trapped in the superconductor [GPS+14].

During a time-dependent variation of the magnetic field (like an AC cycles), there is a movement of magnetic flux inside the superconductor. On the microscopic scale, the vortices

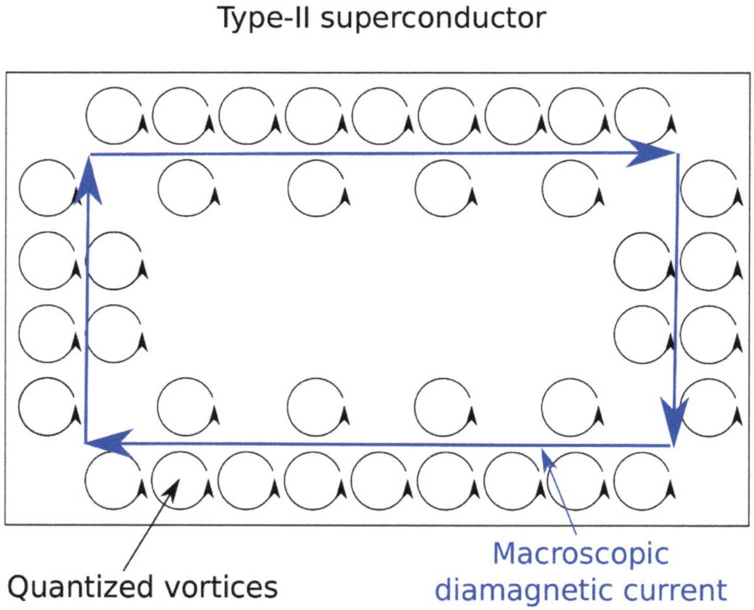

Figure 2.6: Two-dimensional cross-section of an infinitely long rectangular bar of hard superconductor, with schematic illustration of the currents induced by a magnetic field applied along the length of the bar (out of the page). Adapted from [Rey15].

have to move to follow the excitation, in this case given by the time-varying magnetic field. This causes dissipation, which is called the hysteresis loss in hard superconductors. This dissipation is of great practical importance, because many superconducting applications operate in situations where there is a time-varying magnetic field (produced by an external source and/or by the current flowing in the superconductor). AC power applications are a typical case.

2.1.4 High Temperature Superconductors (HTS)

For long time it was thought that, according to the BCS theory, 30 K marked the limit for the existence of superconductivity. Things changed dramatically in 1986, when Bednorz and Muller discovered the Ba – La – Cu – O system ($Ba_xLa_{5-x}Cu_5O_{5(3-y)}$), for which they were conferred the fastest Nobel Prize ever awarded. The so-called high-temperature superconductors are layered materials characterized by copper oxide planes and have critical temperatures of over 100 K. HTS are generally defined as superconductors with a T_c higher than around 23-30 K (30 K is the upper limit allowed by BCS theory, 23 K is the 1973 record that lasted until copper-oxide materials were discovered). YBCO ($YBa_2Cu_3O_7$) is the most famous of these HTS and was discovered by Paul C. W. Chu and M.-K. Wu in 1986 and 1987, respectively. It was the first superconducting material above 77 K, the boiling point of liquid nitrogen, and has consequently paved the way for a much broader range of practical applications. All other materials discovered before this became superconducting at temperatures near the boiling points of liquid helium or hydrogen (4.2 K and 20 K, respectively), which are both more expensive and difficult to obtain than liquid nitrogen. The crystal structure of YBCO is highly anisotropic, with much higher conductivity within the CuO_2 than perpendicular to the planes. Thus, super-currents flow only within the CuO_2 (*a-b*) planes, meaning the trapped field generated by these super-currents is directed along the *c*-axis.

The passing between the mixed state and the normal (non-superconducting) state is not as sharp as it appears in fig. 2.4(b), due to thermal effects. Due to thermal agitation, some vortices can unpin from the pinning centers, move under the action of the Lorentz force and cause some dissipation. The result is that a small finite resistance is observed even at sub-critical currents: for example, in the current-voltage characteristics, the transition to the normal state is relatively smooth. This phenomenon is called flux creep and mainly affects high-temperature superconductors, due to the relatively high operating temperatures [Rey15]. Above a certain temperature flux creep raises to extreme levels and the flux line lattice "melts", causing a transition into a liquid-like state often called the vortex liquid. In this state, magnetic irreversibility is no longer possible and critical current density disappears. This melting temperature depends strongly on the flux density, so that an irreversibility line arises in the *B*-*T* plane and defines the practical range of usability of a superconductor. In some cases, like copper-oxide high-temperature superconductors, the irreversibility field depends dramatically on the orientation of the magnetic field and is considerably lower than the upper critical field.

There are several companies such as American Superconductor (AMSC), SuperPower, SuperOx and THEVA that supply long lengths of REBCO-based HTS tape/wire. The manufacturing techniques differ between them, which results in a different configuration for the final product. It is only in recent years that long lengths of wire has been available commercially, which has made it possible to wind coils and cables for large scale applications, such as electric machines, superconducting magnetic energy storage (SMES) systems and transformers. AMSC's approach to manufacturing REBCO-based HTS wire is based on the RABiTS/MOD (rolling assisted bi-axially textured substrate/metalorganic deposition) technology. The buffer layers (a 75 nm Y_2O_3 seed layer, a 75 nm YSZ barrier layer and a 75 nm CeO_2 cap layer) are deposited by high-rate reactive sputtering onto a metal alloy ($Ni-W$) substrate, and the rare earth doped YBCO is coated onto the buffered substrate. The REBCO is capped with an Ag layer, then oxygenated, and laminated between two metallic stabilizer strips, currently either brass or copper. SuperPower's approach is based on the IBAD/MOCVD (ion beam assisted deposition/metal organic chemical vapour deposition) technology, which involves sputtering a stack of buffer layers to introduce the bi-axial texture for the YBCO layer, which is deposited using MOCVD. A thin cover of silver is then sputtered to provide electrical contact. Depending on the application, this is then electroplated to completely surround the wire. These two kinds of REBCO-based HTS wire are referred to in this manuscript and the samples used for testing are from SuperOx and SuperPower wire. In this work, the terms wire, tape and coated conductor are used interchangeably. The large aspect ratio of the tape and its crystalline structure makes this type of superconductor highly anisotropic and the tape performance is affected greatly by magnetic fields perpendicular to the tape's wide face, i.e. perpendicular to the *a-b* plane.

2.2 AC Loss Mechanisms in Superconductors

Superconductors do not exhibit dissipation when they carry DC current and below their critical parameters, but when they carry AC current and/or are subjected to AC magnetic field they do.

Type-II superconductors consist of several materials, including metallic and sometimes magnetic parts, which lead to different types of AC losses.

2.2.1 AC Loss Contributions in Coated Conductors Tapes and Cables

When a superconductor tape or cable is subjected to time-varying magnetic field, dissipative phenomena take place. Different types of losses can be distinguished, as follows:

Hysteresis Loss

The hysteresis losses occur in the superconducting material and are caused by any time-varying magnetic field which makes the fluxons move. When the magnetic field changes, the flux-line pattern and internal magnetic field change as well. The magnetic-field variation inside the material induces a voltage, which creates dissipation because of the normal conducting regions associated with each vortex. The time-varying magnetic field can be externally applied (magnetization losses) or the self-field associated with a transport current (transport losses). Despite the different origin of the magnetic field, the physical mechanism causing dissipation is the same.

Eddy Current Loss

When an external time-varying magnetic field penetrates into a normal conductor such as the Hastelloy, the silver and the copper, it induces an electric field that makes currents flow. These are known as eddy currents.

Magnetic Loss

Magnetic losses are the losses occurring in the magnetic parts (if present, usually for manufacturing reasons), when they are subjected to varying magnetic fields. The presence of magnetic materials is not necessarily negative. For example they can be used to produce shields for the superconductor and reduce its losses [Kru14]: in that case the trade-off between the reduced losses in the superconductor and the additional losses in the magnetic material must be assessed.

2.2.2 Solutions to Reduce AC Losses

There are several strategies for reducing the AC losses in superconductors, more specifically the magnetization losses caused by an external magnetic field. Hysteresis losses in REBCO conductors can be reduced by striating the tape in filaments, but if these latter are coupled, the positive effect of striation is nullified and another loss contribution, the coupling loss, arises. The basic approach to reduce the eddy current losses is to increase the effective resistivity of the material. The idea of shielding the tape also helps to reduce the magnetic losses.

Filaments in HTS Coated Conductors

Generally speaking, it is convenient to have small superconducting filaments rather than a large superconductor. For example, the magnetization losses of a thin strip are proportional to the square of the width of the strip (see eq. 2.3 later). This means that if the strip is subdivided into N narrower strips, the loss of the tape are N times lower than those of the original tape (the losses of each narrow strip are N^2 times lower, but there are N of such strips). So, manufacturing superconductors in the form of this filaments is a good approach to reduce the losses.

One should note, however, that in real applications the superconductor tape is very long, the transverse resistance between the filaments is consequently very low, and it is therefore extremely easy for the current to go from one filament to the other. When currents flow from one filament to another, they can couple the filaments together into a single large magnetic system and cause additional dissipation because of the resistive material they flow through. Then, if the filaments are coupled, they behave as a sole unstriated superconductor and the advantage of having narrower filaments is canceled. Increasing the transverse resistance between the filaments (e.g. with well defined grooves of insulating material between them) can help to limit the flow of current in the resistive material between the filaments.

Subdividing the superconductors into filaments can be a problem in REBCO conductors. Striation is discussed in detail in Chapter 5 when discussing AC loss mitigation.

Transposition

Superconductors are always placed in devices such as cables or coils, which have metal terminations at their ends, where the current is injected. Even under the assumption that the filaments are perfectly isolated from each other, they can easily become coupled through those terminations, in this way canceling the positive effect of having narrow filaments.

A practical way of making filaments uncoupled is by twisting them, so that the current paths are transposed: each path changes places with every other stands along the length of the conductor. The principle is illustrated in Fig. 2.7, where the voltage inducing the current loops inverts its direction every half twist. The length available for coupling is reduced: it is no longer given by the total length of the strands, but by half of the so-called "pitch length".

One can produce wires with twisted filaments with LTS, M_gB_2, and also Bi-2223 superconducting wires, but it is much more difficult in the case of REBCO coated conductors, due to their geometry and the extremely small thickness of the superconductor layer. A solution is to obtain the transposition of the current paths using to geometry of a cable consisting of multiple coated conductors. For example by winding the coated conductor tape around a cylindrical former. This is the idea behind the <u>C</u>onductor <u>O</u>n <u>R</u>ound <u>C</u>ore (CORC®) design.

Another geometry offering transposition of the current paths is the <u>R</u>oebel <u>A</u>ssembled <u>C</u>oated <u>C</u>onductor (RACC) cable, which is manufactured by punching HTS coated conductor into meander-shaped strands and assembling them in a kind of braid structure. One can also think of introducing filaments in the strands composing CORC® or RACC cables. However, in a CORC® cable, the filaments are fully transposed only at the level of the individual layer (the filaments do not change place with each other between different layers). In addition, one should also assume no gap between the tapes (with filaments), because the gap makes the filaments situated at the edge of a tape electromagnetically different from those situated at its interior. In a RACC cable, the transposition of the filaments is only partial, because the filaments do not interchange all the positions: for example, a filament situated on one side of the cable never reaches the other side.

Magnetic Shielding

Another way of reducing AC magnetization losses is by magnetic shielding. The basic principle is to try to pull magnetic field out of the superconductors with the help of shields of ferromagnetic materials, properly shaped and positioned near the most critical part of the superconductor, for example the edges of an HTS coated conductor. When a time-varying magnetic field (perpendicular to the superconductor) is present, the magnetic flux lines are taken away from the superconductor and concentrated in the shield. The magnetization losses of the superconductor material are lower (because of the much lower magnetic field impinging the superconductor), but there are additional magnetic losses in the shielding. However, depending on the shape and size of the shield as well as on the properties of the magnetic material, the reduction of the losses in the superconductor can be larger than the additional losses generated in the shield.

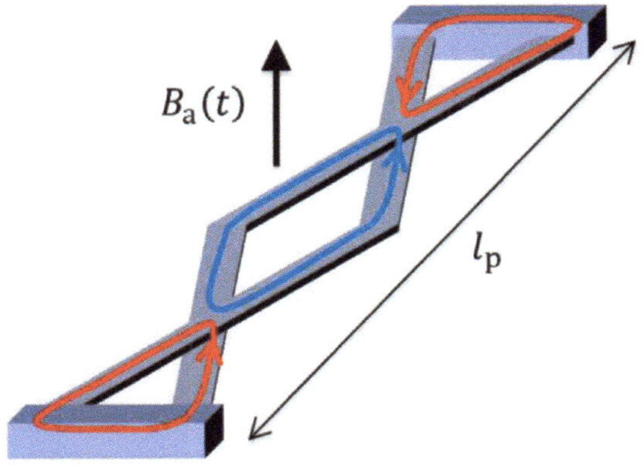

Figure 2.7: Current loops induced in a conductor composed of two transposed current paths.

2.2.3 Analytical Formulas for AC Losses in Tapes and Cables

The following analytical formulas are for calculating AC losses in situations and geometries relevant for this thesis. A more complete discussion on analytical models for AC losses in superconductors can be found in [MMWS13].

Magnetization Losses of a Thin Superconductor Strip

Halse [Hal70], Brandt and Indenbom [BI93], and Zeldov et al. [ZCMD94] calculated the current and field profiles in a thin superconductor strip in a perpendicular magnetic field. They developed analytical equations for calculating the hysteresis losses in a thin superconductor strip on the basis of the critical state model. For a strip of width $2a$, thickness d and critical current I_c in a perpendicular magnetic field H_a and frequency f, the hysteresis loss per unit length is as follows:[4]

$$Q_{\mathrm{Brandt}} = 4\pi\mu_0 a^2 J_c H_a \left[\frac{2H_c}{H_a}\ln\cosh\left(\frac{H_a}{H_c}\right) - \tanh\left(\frac{H_a}{H_c}\right)\right] \tag{2.3}$$

where $H_c = J_c/\pi$, $J_c = j_c d = I_c/2a$.

Magnetization Losses of an Array of Thin Superconducting Strips

At sufficiently large applied fields, the hysteresis loss in the superconductor is expected to reduce proportionally to the number of filaments. Mawatari [Maw96] provided an analytic formula for the loss of an infinite array of superconducting strips subjected to perpendicular field, a geometry that approximates well a striated coated conductor with electrically uncoupled filaments (introducing an infinite resistance between filaments).

The hysteresis loss per unit length per cycle of a striated superconductor strip made with n_{fil} number of filaments in an applied field H_a is given by:

$$Q_{\mathrm{Mawatari}} = -n_{\mathrm{fil}}\frac{h_0^2 Q_c L^2}{\pi^2 w^2}\int_0^1 (1-2s)\ln\left[1-\frac{\sin^2\left(\frac{\pi w}{L}\right)}{\cosh^2(h_0 s)}\right] ds \tag{2.4}$$

where $h_0 = \pi H_a/J_c$, $Q_c = \mu_0 I_c^2/\pi$, $J_c = j_c d = I_c/2w$ and $L = 2w + w_{\mathrm{groove}}$. w is the width of a filament, w_{groove} the width of the space between two filaments and d the superconductor thickness.

[4]Due to the different utilized notation and approach, the expression for the losses has a different form in the papers mentioned above, but they are in fact equivalent. Here the one from [BI93] is used.

Eddy Current Losses

The eddy current losses, which are expected to occur in the normal metal parts of the tape, such as the Hastelloy, the copper layers and the silver layers, of a normal conductor slab are given per unit volume by [Mül97]:

$$P_{\mathrm{ec}} = \frac{\pi^2}{6} \frac{B_{\mathrm{a}}^2 f^2 W_{\mathrm{nc}}^2}{\rho_{\mathrm{nc}}} \tag{2.5}$$

where B_{a}, W_{nc} and ρ_{nc} represent the magnetic field amplitude, the width and the resistivity of the normal conductor, respectively.

Transport Current Losses

Norris [Nor70] proposed an analytical method to estimate the transport losses of a superconductor strip. For a superconducting strip carrying a transport current with magnitude I_{a}, the AC transport losses per cycle per unit length can be calculated as

$$Q_{\mathrm{Norris,strip}} = \frac{\mu_0 I_{\mathrm{c}}^2}{\pi} \left[\left(1 - \frac{I_{\mathrm{a}}}{I_{\mathrm{c}}}\right) \ln\left(1 - \frac{I_{\mathrm{a}}}{I_{\mathrm{c}}}\right) + \left(1 + \frac{I_{\mathrm{a}}}{I_{\mathrm{c}}}\right) \ln\left(1 + \frac{I_{\mathrm{a}}}{I_{\mathrm{c}}}\right) - \frac{I_{\mathrm{a}}}{I_{\mathrm{c}}}^2 \right] \tag{2.6}$$

Norris' formula can be used for calculating the AC transport losses of cables with rectangular cross-section shape such as RACC cables. A second formula is provided for the losses of a superconductor with elliptical cross-section.

Transport Losses of a Hollow Superconductor Cylinder

The monoblock model is an analytical model developed for calculating the transport losses of a superconducting solid tube [VM95]. It can be used for getting an approximate estimation of the transport losses of CORC® cables. However, the model does not account for the gaps between the tapes that occur in a CORC® cable: in these gaps there can be a significant penetration of magnetic flux perpendicular to the flat face of the tapes, which can significantly contribute to the losses of a CORC® cable.

The expression of monoblock model in loss per unit length and per cycle is given as follows [VM95]:

$$Q_{\mathrm{monoblock}} = \frac{\mu_0}{2\pi} \frac{I_{\mathrm{c}}^2}{h^2} \{(2 - Fh)Fh + 2(1 - Fh)\ln(1 - Fh)\} \tag{2.7}$$

with $h = (D_{\mathrm{c}}^2 - D_{\mathrm{f}}^2)/D_{\mathrm{c}}^2$ and $F = I_{\mathrm{a}}/I_{\mathrm{c}}$. D_{c} is the outer diameter of the cable, D_{f} the outer diameter of the former, I_{a} the transport current amplitude of the cable and I_{c} the critical current of the cable.

The analytical formulas by Brandt and Mawatari are used for estimating the AC magnetization losses of unstriated and striated tapes, respectively. The formulas by Norris and of the monoblock model are used for estimating the AC transport current losses of RACC and CORC® cables, respectively.

3 Preparation of the Samples and Experimental Setups

This chapter is divided in two sections, each describing the preparation and the experimental setup for characterizing the properties of HTS tapes and cables, respectively.

Among the various manufacturers in the world, for example American Superconductor [Supa], Fujikura [Fuj], Superconductor Technologies Incorporation [STI], SuNAM [SuN], SuperOx [Supb] SuperPower [Supc] and THEVA [THE], that commercially produce REBCO tapes, two were selected for preparing test samples. The first one, SuperPower, for the reproducible and good homogeneity and high critical current; the second, SuperOx, for the possibility of customizing the tapes such as the thickness of the copper and silver layer.

The division of REBCO tapes into narrow strips as a method to reduce the AC magnetization losses was proposed by Carr and Oberly [CO99] in 1999 and its effectiveness was demonstrated with various techniques such as chemical etching [MZX+09, TC07], mechanical scratching [KMS13, PKC+02] or laser ablation [AKY+04, AYK+05, CBH+02, DVK+14, DGK+15, KLHS13, KLC+15, LBA+05a, LBA+05b, LBA07, LBK+06, LMH+13, MGC+05, NVD+14, PUJ+06, PKM+07, SCB05, SLC+03]. A review on the topic can be found in [GK16]. The laser ablation technique was selected for this work for striating the tapes due to the making of narrow laser beams, 20 µm, the preciseness of the laser cut, the flexibility in parameters choice such as the power, the pulse energy, and the frequency, which allows optimizing the tape architecture and the possibility of making a high number of filaments. It has been shown that it is possible to introduce up to 120 filaments in a 12 mm wide tape with the laser ablation technique [DVK+14, DGK+15, NVD+14].

The reduction of the AC losses in a cable requires that the individual strands be transposed. Among various HTS cables, the Conductor On Round Core (CORC®) [GŠVS15, MSCvdL14, SŠG14, vdL09, vdLNM+13, vdLWN+16, VKR+15] and Roebel Assembled Coated Conductor (RACC) cables were chosen because of their complete strand transposition and their high engineering current density [GNK+06, GFH+07, GFK+09, GGP+14, GVKZ16, TVG+11, VGT+11].

CORC® cable was proposed and manufactured by D. C. van der Laan in 2009 [vdL09] and it is currently produced by Advanced Conductor Technologies LLC[5] [Tec]. RACC cable was first proposed and developed by W. Goldacker et al. in 2006 [GNK$^+$06] and it is currently being developed at the Institute for Technical Physic (ITEP)[6] of the Karlsruhe Institute of Technology (KIT), at Industrial Research Limited (IRL) and at SuperOx.

A direct comparison between CORC® and RACC cables is difficult due to their different architecture (round and rectangular cross section, respectively) and the tape manufacturer (SuperPower and SuperOx, respectively). That is why in this work, cables with similar critical currents were assembled, measured and compared. Striated cables are also introduced in order to observe the effectiveness of striations to reduce the AC losses.

3.1 2G-HTS Tape Preparation and Experimental Setup

As previously said, the most common process to reduce the magnetization losses is through a filamentarized structure. Its successful application to tapes with thick copper stabilization (20 µm), where power applications require stabilizer, is not straightforward [KLHS13, KMS13, LMH$^+$13], due to the difficulty of controlling the quality of the groove and thus avoiding filament coupling [PUJ$^+$06, DUT$^+$07, MZX$^+$09, DVK$^+$14, KKL$^+$16].

In a previous work of our group [DGK$^+$15], significant AC magnetization loss reduction was observed for striated tapes with 5 µm of copper. In that work, the influence of the way of depositing copper and of its thickness was investigated. For this, several HTS samples with

[5] http://advancedconductor.com/

[6] https://www.itep.kit.edu/english/index.php

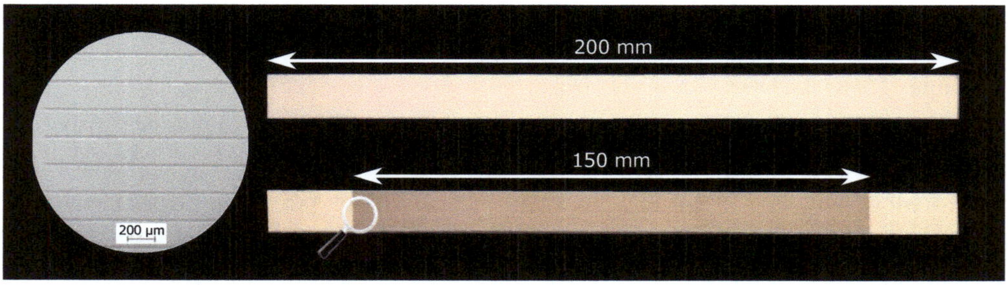

Figure 3.1: Top: reference 200 mm long unstriated sample. Bottom: 60-filament tape striated over a 150 mm length. A magnified view of the striation is also shown on the left, which is a Scanning Electron Microscopy (SEM).

copper stabilization of 5 µm and 10 µm were prepared. Tapes with 4 mm and 12 mm width were provided by the manufacturers SuperPower and SuperOx, respectively.

Two manufacturing processes were used to make the striations on the samples. The striations were produced either before or after deposition of the copper layer. Those processes are identified throughout this thesis as Striated Before Electroplating (SBE) and Striated After Electroplating (SAE).

For the striated tapes, 150 mm long striations were made in the central part of 200 mm long samples, as shown in Fig. 3.1. The unstriated ends were used to safely inject the current for critical current measurements. For AC magnetization loss measurements, all the tapes, including the reference non-striated ones, were cut to a length of 150 mm. The striated tapes had therefore open-end filaments.

The filaments were created by means of a laser TRUMPF TruMicro 5025. This laser has an infrared (IR) wavelength of 1030 nm, a pulse frequency of 400 kHz with pulse duration less than 10 ps and a maximum power of 25 W with maximum pulse energy of 125 µJ. More details can be found in [NVD+14]. All the striated tapes were scribed using 20 % of the laser power and their striations were performed down to the level of the Hastelloy, which is 60 µm thick.

Scanning Electron Microscopy (SEM) was carried out to examine the quality of the laser groove. SEM itself allows observing the top of the groove and, once coupled to Focused Ion Beam (FIB) removing the material of a small zone from one of the groove, observing the cross section of it [RP01].

3.1.1 Preparation of REBCO Tapes

Various tape batches for AC magnetization loss analysis were manufactured. Each batch consists of one reference non-striated tape and striated tapes with various numbers of filaments, one or two copper thicknesses and one or two striation processes. Both striation processes, SBE and SAE, are schematically illustrated in Figs. 3.2(a) and 3.2(b), respectively.

Tapes using the SBE process are tapes with only silver cap which were first scribed by means of the laser, then oxidized [KLC+15], and finally electroplated with copper. Tapes using the SAE process were first electroplated with copper and then striated with the laser. SBE tapes have a similar structure to SAE but they do not have a copper layer at the bottom of the tape. For both striation processes, the groove width was about 20 µm and the expected thickness of the electroplated copper was 5 µm or 10 µm.

In order to obtain the best groove quality, the striation processes were optimized: various power levels as well as numbers of runs of the laser were tested and the results verified by using Scanning Electron Microscopy (SEM).

Four batches were prepared from tapes manufactured by SuperOx. Two batches were electroplated with SAE process with 5 µm and 10 µm thick copper stabilization and the other two with SBE process with 5 µm and 10 µm of copper as well. The entire set of SBE tapes were striated by applying 12 runs for each groove. The SAE tapes with 5 µm and 10 µm of copper were striated by applying 50 and 60 runs for each groove, respectively. The aim of the comparison of different tapes was to observe the influence of the copper thickness and of the striation process on the AC magnetization losses. The results are shown and discussed in chapter 5.

Tapes from the manufacturer SuperPower with a width of 4 mm and with 5 µm of copper stabilization were prepared. Tapes with 3 filaments, 5 filaments, 7 filaments, 10 filaments, 20 filaments and 30 filaments were striated with the SAE process. Twenty-five runs per groove were performed in order to make the striations. The aim of the comparison between tapes with different number of filaments was to find the best candidate for assembling CORC® cables. This comparison was also used for the analysis of the influence of different number of filaments on the AC losses for the same striation process and thickness of copper stabilization. The results are shown and discussed in chapter 6.

Table 3.1 summarizes the main information of the prepared tapes such as the width, the striation process, the thickness of copper stabilization and the self-field critical current at 77 K.

In order to prevent the coupling between the copper on the top and either the copper or the Hastelloy on the bottom across the edges of the conductor, two laser lines were scribed as close as possible to the conductor's edges along the length.

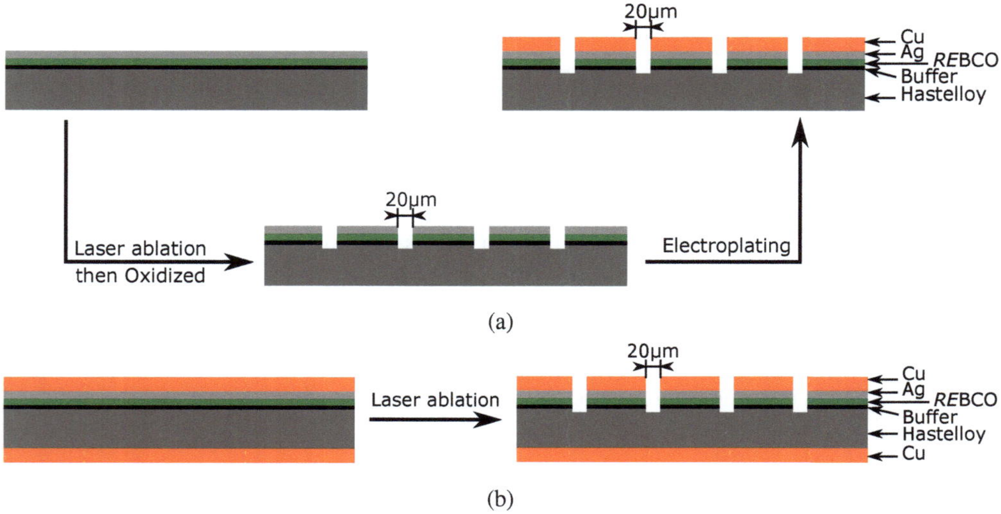

Figure 3.2: Schematic illustration of the cross section of the two striation processes: (a) Striation Before Electroplating (SBE) process. (b) Striation After Electroplating (SAE) process.

Table 3.1: Prepared superconducting tapes for AC loss measurements. The width, the striation process, the thickness of the copper and the critical current of each tape is referred. The non-striated striated tapes are labeled as Ref.

Width (mm)	Striation Process	Copper Thickness (µm)	Number of Filaments (Critical Current in A)
12*	SAE	5	Ref (345), 10 (330), 20 (305) 40 (255) and 60 (205)
		10	Ref (335), 10 (305), 20 (270) 40 (215) and 60 (195)
	SBE	5	Ref (350), 10 (300), 20 (325) 40 (285) and 60 (225)
		10	Ref (370), 10 (325), 20 (280) 40 (215) and 60 (190)
4**	SAE	5	Ref (105), 3 (102), 5 (98), 7 (95) 10 (91), 20 (66) and 30 (47)

*Sample's ID: SOx-KIT

**Sample's ID: SP-KIT-20140429-2b

3.1.2 Calibration-free Method Setup

The measurements of the AC magnetization losses of HTS tapes were performed by means of the calibration free method [ŠGV05] at 77 K. This setup was designed and built by M. Vojenciak at the ITEP.

The method is based on the observation that the losses in the sample constitute a fraction of the power supplied to the whole system by the AC source; this fraction is generally small, but can be detected by building a symmetric system consisting of two pick-up coils, subjected to the same external uniform magnetic field (produced by two sets of power-supplying coils), connected in series, but wound in opposite directions. In the absence of sample, the system is perfectly balanced and the voltage measured by the connected pick-up coils is - ideally - zero. When the sample is inserted inside one of the measuring coils, the symmetry is broken, the total voltage induced in the two pick-up coils by the external field cancels out because of their opposite winding directions, and the measured signal gives directly the loss of the sample. This method has the big advantage that the same experimental set-up can be used to measure samples of different shapes and sizes, without the need of building *ad hoc* pick-up coils wound around the samples.

Figure 3.3 shows the sketch of the setup for the AC loss measurement, which consists of two power-supplying coils (A and B) connected in series in order to create two ideally identical AC magnetic fields. These coils are composed of four copper racetrack coils with curved ends. By bending the racetrack coil ends, the four racetrack coils can be placed closer to each other, which increases the produced magnetic field while keeping an entrance large enough to introduce the sample as shown in Fig. 3.4.

Two pick-up coils C and D, connected in series but in opposite direction, ensure the voltage balance by subtracting the voltage of the first coil C by the second coil D. They are formed by non-connected strands of the power-supplying coils A and B.

The coil-powering current I_a is supplied by a power amplifier controlled by a waveform generator and it is measured by a Rogowski coil that is connected to the first input of the lock-in amplifier (model SR 830 DSP) as phase reference. The Rogowski coil measures the magnetic field generated by the power cable (sinusoidal current) and it supplies a voltage proportional to the current and shifted by 90°.

In practice, the signal from the pick-up coils (C and D) V_b is not equal to zero due to imperfections of the power coils (A and B). It therefore needs to be compensated. This signal is connected to the second input of the lock-in amplifier. The signal V_b is composed of two components, one that is in phase with the current (90° shift, output of the Rogowski coil) and another that is out of phase (sinusoidal component), which is used to calculate the losses:

$$V_b = \sum_{k=1}^{n} A_k \cos(\omega t) + \sum_{k=1}^{n} B_k \sin(\omega t) \tag{3.1}$$

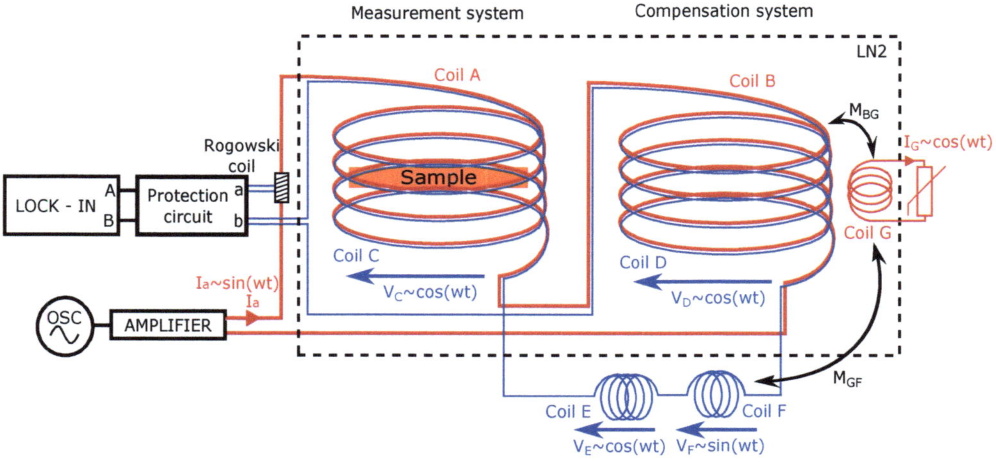

Figure 3.3: Diagram of the calibration-free method setup. The principle of the location of the hardware and the components are shown.

The lock-in amplifier has the capability of separating the two components (in phase and out of phase) of a signal. The sensitivity of the lock-in amplifier is tuned from the highest voltage of both components. This means that, if the voltage of one of them is much higher than the other one, then the smallest voltage is not readable. That is why, two additional coils E and G were introduced to compensate each component.

The coil E is connected in series with the coils C and D and it is located out of the liquid nitrogen close to the power coil B in order to catch its magnetic flux. By adjusting the position of the coil E, the amount of the caught flux sets the voltage level in order to lower or rise the in-phase component.

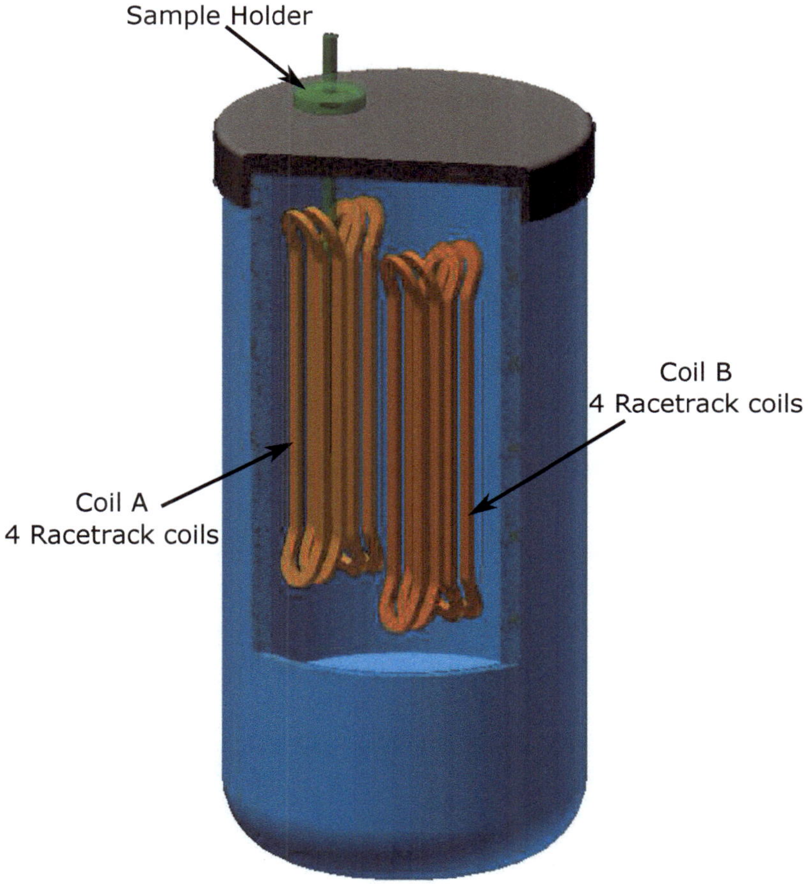

Figure 3.4: Schematic illustration of the setup for measuring magnetization AC losses with the calibration-free method.

To reduce or increase the out of phase component, a variable resistance connected in series with the coil G (located next to the coil B) is tuned so as to create a cosinusoidal current due to derivation of the carried current of the power coil B with the mutual inductance M_{BG}. This circuit (the coil G and the variable resistance) creates a magnetic flux that is caught by the coil D. Due to the mutual inductance between the coil G and D (M_{GD}), and the current carried by the coil G, a voltage out of phase is created into the circuit consisting of the coils C, D and E. This voltage is represented by a coil F in Fig. 3.3

Firstly, the coils E and G are tuned in order to get the signal V_{b} without sample (background signal) as low as possible. Then the sample is inserted and the voltage signal with sample is measured. The sample, which can have a maximum length of 150 mm (see Fig. 3.5), is inserted into the power coil A. Its orientation is perpendicular to the produced field, because this is the orientation for which the losses are most significant.

The magnetization loss of the samples is calculated with the effective voltage that comes from the measurement with sample after subtracting the background signal (without sample) and the current that comes from the Rogowski coil:

$$P_{\mathrm{sample}} = V_{\mathrm{b,RMS}} . I_{\mathrm{a,RMS}} \tag{3.2}$$

where $V_{\mathrm{b,RMS}}$ is the effective value of the quadrature component of the voltage V_{b} (sinusoidal component) and $I_{\mathrm{a,RMS}}$ is the effective value of the current I_{a}, which is measured by the Rogowski coil.

3.2 2G-HTS Cable Preparation and Experimental Setup

CORC® and RACC cables are assembled with tapes and strands, respectively, in specific configurations to obtain a conductor with a high current carrying capability and sufficient tensile strength. The geometry of CORC® and RACC cables allows having – at least in theory – a uniform current distribution between the tapes and the strands. CORC® and RACC cables

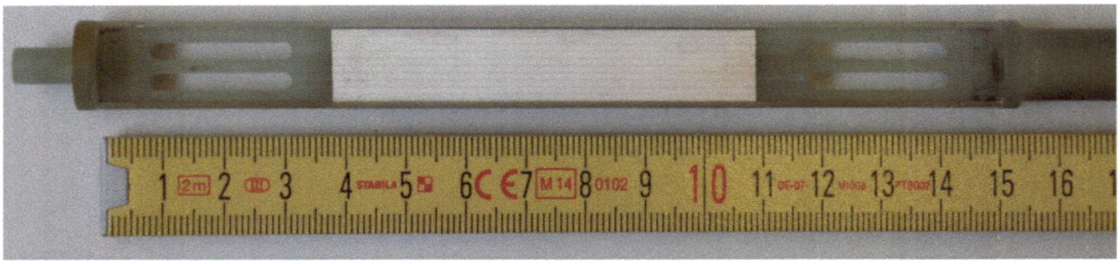

Figure 3.5: Sample holder with striated 70 mm long and 12 mm wide tape.

with striated conductors were also assembled in order to test the effectiveness of striations on the AC losses.

A CORC® cable is assembled by winding several layers of 4 mm wide tapes around a cylindrical former. A RACC cable consists of several meander-shaped 5.5 mm wide intertwined strands. The number of tapes or strands constituting the CORC® or RACC cable defines the current-carrying capacity of the cable. Both, tapes and strands, were electroplated with similar copper thickness for stabilization (5 µm), but using samples from the manufacturer SuperPower and SuperOx, respectively.

A direct comparison between CORC® and RACC cables is difficult due to their different architecture and the tape manufacturer. That is why in this work, cables with similar critical currents were assembled and their preparation are shown in section 3.2.1. Their AC losses were measured and are investigated in chapter 6.

In a CORC® cable, transposition occurs within layers for both type of tapes (striated and non-striated tapes), assuming no gap between the strands. In a RACC cable, the non-striated strands are fully transposed, but the filaments are not: a filament situated on one side of the cable never reaches the other side.

The critical current I_c of the cable determines the number of tapes for a CORC® cable or the number of strands for a RACC cable. This critical current can be estimated as the sum of the single tape or strand critical current reduced by an amount due to the self-field, which can be roughly estimated as 20-30 % for both cables. This reduction depends on the number of layers for the CORC® cable or on the number of strands for the RACC cable as shown in [vdLNM+13] and [GGP+14], respectively.

3.2.1 Preparation of High Current HTS Cables

In total, six different CORC® cables were manufactured by Advanced Conductor Technologies LLC with the main data given in Table 3.2.

Three CORC® cables were made with non-striated tapes (label "C_R") and three with striated tapes (label "C_S"). All tapes have the same copper thickness (5 µm), the same tape width (4 mm) and are from the same manufacturer (SuperPower). The number of layers was varied from 2 to 4. The striated tapes have 5 filaments because it represents a good compromise between AC loss reduction and preservation of the critical current [VKR+15]. The superconducting tapes were wound on a tube and the layers are wound in opposite direction as shown in Fig. 3.6.

The ends of the cable were inserted and soldered into copper tubes. Two contacts ("L_{C0}" and "L_{C1}" in Fig. 3.7(a)) were soldered inside the copper current leads in two different locations. Three contacts ("T_{C1}", "T_{C2}" and "T_{C3}" in Fig. 3.7(a)) were soldered on the three

Table 3.2: Prepared CORC® cables with their corresponding critical current. The ones label as "C_R" correspond to the CORC® cables with non-striated tapes and the ones label as "C_S" with striated tapes. The number following "C_R" or 'C_S" represents the number of strands of the cable.

Reference cables (non-striated)	I_c (A)	Striated cables (5 filaments)	I_c (A)
C_{R6} (3 tapes x 2 layers)	454	C_{S6} (3 tapes x 2 layers)	410
C_{R9} (3 tapes x 3 layers)	795	C_{S9} (3 tapes x 3 layers)	588
C_{R12} (3 tapes x 4 layers)	998	C_{S12} (3 tapes x 4 layers)	963

Sample's ID: SP-KIT-20140429-2b

tapes located on the outer layer at a distance of eight transpositions lengths (144 mm). Those additional contacts are added to monitor the homogeneity of voltages.

The striations of strands were done over 150 mm length using the SAE process, leaving 200 mm length for contacts. The filaments were prepared at ITEP using 20 % of the laser power with 50 runs for each grooves. The striated CORC® cables strands were wound inwards.

The manufacturing of a RACC strand starts from a single 12 mm wide tape shaped by a dedicated mechanical punching machine into a meander-shaped strand. The mechanical punching

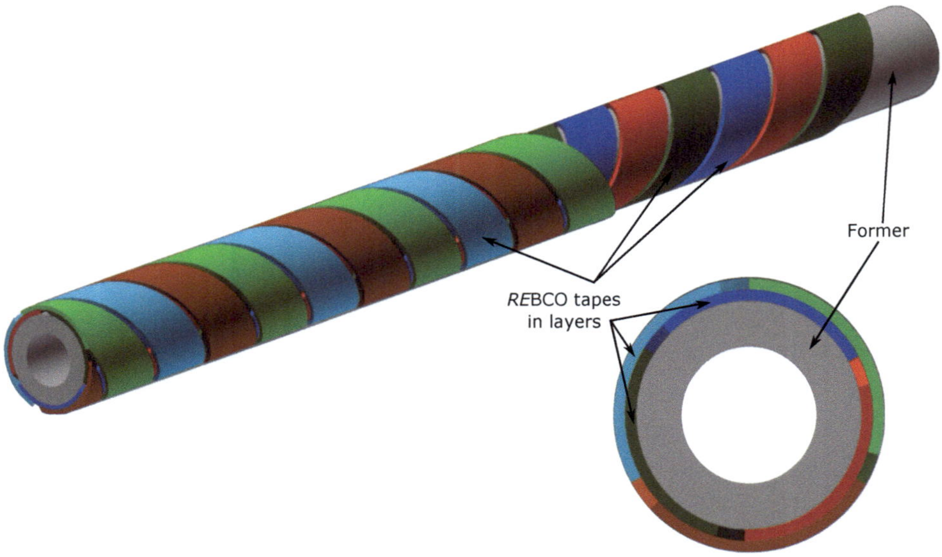

Figure 3.6: Schematic illustration of a CORC® cable with 2 layers and three tapes per layer. A cross section of the cables is also shown.

process is used by KIT, IRL and SuperOx and can achieve a precision lower than 50 μm for the strand width [GGP+14]. A schematic illustration of the punching process, which is used for manufacturing the RACC strands is given in Fig. 3.8(a). The machine has two punching dies, one for the top and another one for the bottom of the tape width. The machine cuts the tape several times along a transposition length. This number depends on the transposition length. Two times punching was used for the configuration in this work. The first punching is always the same size as the die but the other one changes according to the transposition length.

Various transposition lengths can be made: a length of 125.8 mm, 226 mm or 426 mm for tapes with a width of 12 mm. Once all the strands are produced, the RACC cable, for short lengths, is assembled by hand as shown in Figs. 3.8(b) and 3.8(c). A gap between the strands must be introduced in order to prevent the damage of a strand by the neighboring strands. The transposition length is determined by the number of strands, which is itself determined by the desired critical current of the cable, and the gap between them.

Two RACC cables were prepared and assembled by A. Kario [KVK+16] at KIT. Both use SuperOx 12 mm wide punched tapes with 5 μm of copper stabilization (ID: SOx-2015-43). All RACC cables consist of 10 strands with a transposition length of 126 mm and six transposition lengths. After punching a tape with 12 mm width, the new width of the strand is 5.5 mm.

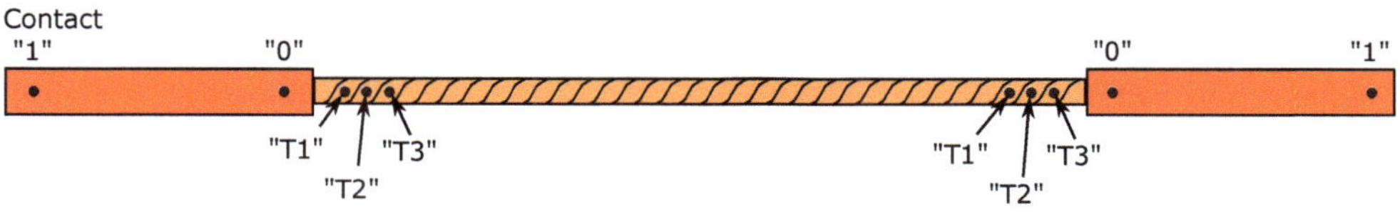

(a)

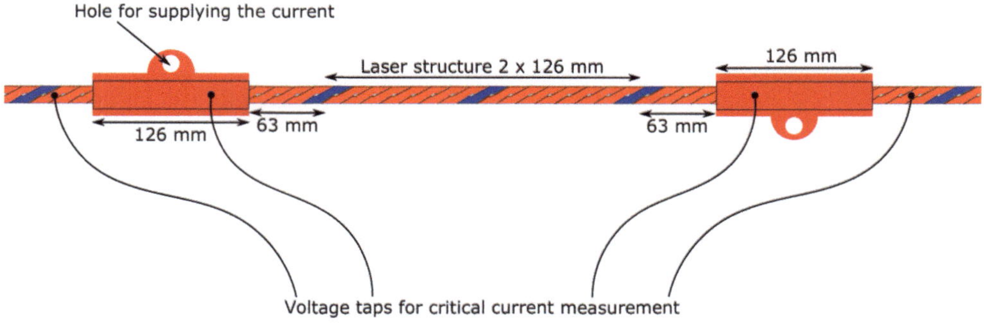

(b)

Figure 3.7: Schematic illustration of the preparation of the striated and reference cables for electrical DC and AC transportation measurements. (a) CORC® cable: location of voltage taps is shown. For striated strand cable, the striation are on the inner side of the layer. (b) RACC cable: the copper current leads (in red) are shown as well as the voltage taps.

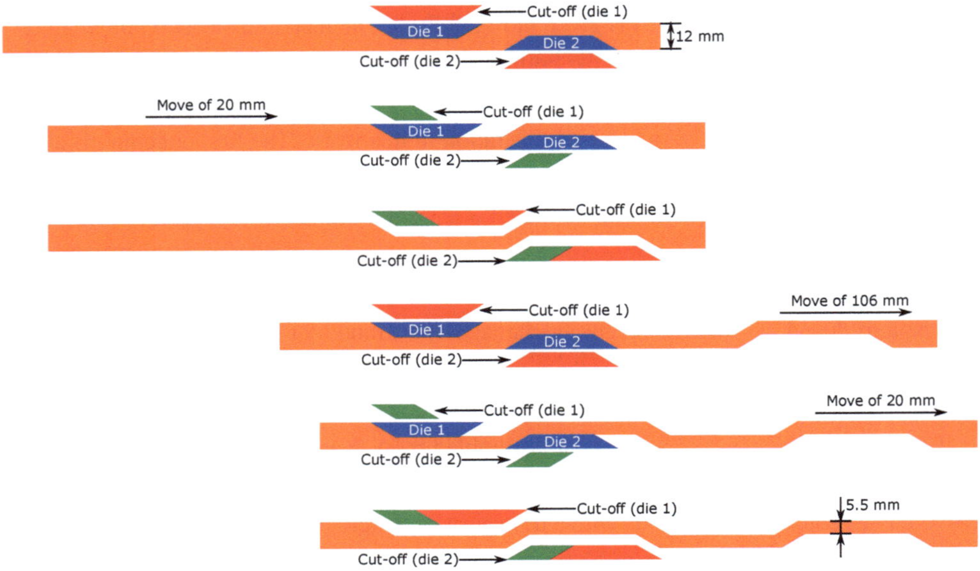

(a)

(b)

(c)

Figure 3.8: (a) Schematic illustration of the punching machine operation for a sample of 12 mm. (b) 3D illustration of the assembly of seven strands for making a RACC cable. (c) Top view of a RACC cable with five strands.

The first RACC cable was assembled with non-striated tapes and it was used as reference. The second RACC cable was assembled with striated tapes along two transposition lengths (252 mm) in the central area of the cable. The five filaments were cut with the SAE process using 20 % of the laser power with 50 runs for each groove. The critical current of the non-striated RACC cable is 914 A and that of the striated RACC cable is 993 A.

For all RACC cables, copper current leads, measuring one transposition length (126 mm), were soldered close to the ends of the cables with PbSn solder at 240 °C. Figure 3.7(b) illustrates the dimensions as well as the location of the copper current leads and of the region with filaments.

3.2.2 Calorimetric Method Setup

The boil-off calorimetric method [MMB$^+$13] is a flexible technique for measuring the AC magnetization losses, the AC transport losses and a combination of the two. It consists in measuring the evaporation of the coolant, which is proportional to the AC losses caused by the

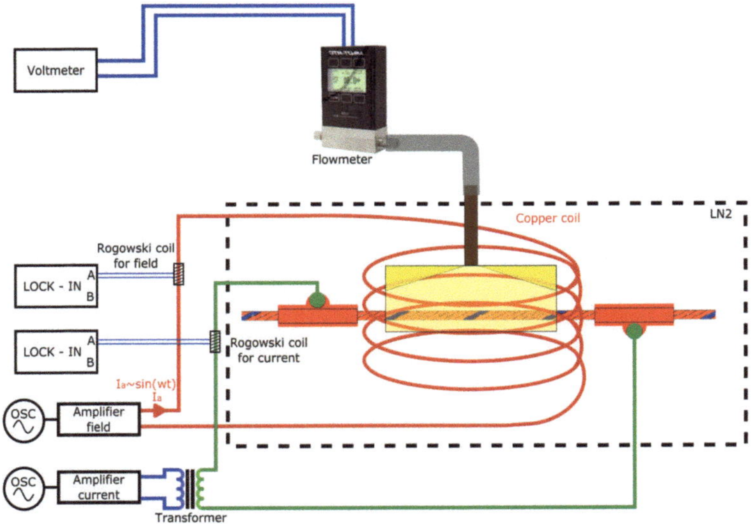

Figure 3.9: Setup illustration of the calorimetric method. The hardware and the components used for measuring the AC losses are shown.

heating of the cable. The proportionality between AC losses and evaporation is determined by means of a calibration process employing a heater of known resistance. All the measurement were performed in liquid nitrogen at atmospheric pressure. The volume of the boiled-off liquid nitrogen resulting from 1 W is 0.0225 L h^{-1} [Eki06] at atmospheric pressure.

For those measurements, the cable is placed inside the uniformity region of the field produced by a racetrack coil. The cable is also connected to a transformer that is connected to a current supply controlled by a waveform generator for supplying the cable with an AC current. The AC losses generated in the superconductor evaporate the liquid nitrogen, which is collected by means of a bubble catcher and then sent to a flowmeter, which is connected to a voltmeter. The bubble catcher box has a length of 126 mm and thus covers one transposition

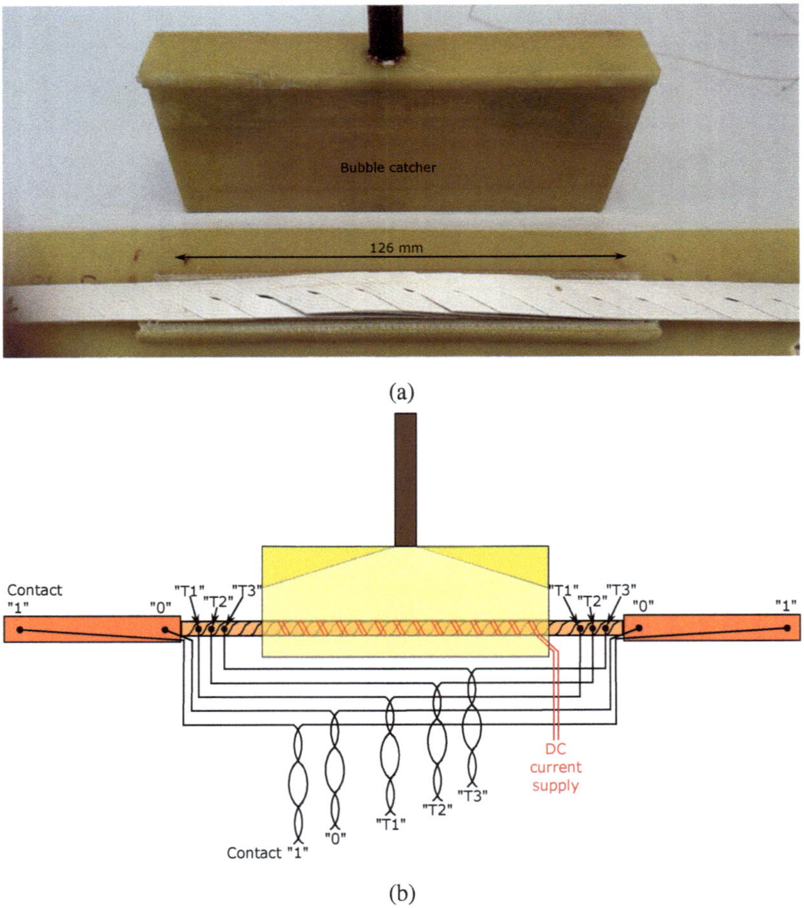

Figure 3.10: (a) Bubble catcher fitted to one transposition of striated RACC cable for the calorimetric method. (b) Schematic illustration of a CORC® cable with bubble catcher, voltage taps and heater.

length of the RACC cable and seven transposition lengths of the CORC® cable. An overview of the setup is shown in Fig. 3.9.

For measuring the AC magnetization losses, the AC current of the racetrack coils was varied in order to apply various AC magnetic fields to the cable. For measuring the AC transportation losses, the HTS cable was directly fed with the AC current. For the measurement of combined transport and magnetization AC losses, both methods were simultaneously applied.

A heater made from a resistive wire (copper nickel alloy with small additions of manganese: CuNi44, 26 Ω) is used to calibrate the measured volume of evaporated nitrogen to the AC losses. The shape of this heater is as close as possible to that of the cable under investigation: a rectangular one for the RACC cable and a cylindrical one for the CORC® cable. Its length corresponds to the bubble catcher box (Fig. 3.10(a)). For the RACC cable, the heater was wound around the edge of a rectangular G-10 former with a empty central area and placed above the cable within the bubble catcher. For the CORC® cable, it was wound around the cable itself as shown in Fig. 3.10(b).

The calibration process consists in supplying a heater with DC current to reach the evaporation level of the different applied magnetic fields (left part of the plot in Fig. 3.11). For each evaporation level (corresponding to a different applied field) of the heater, the voltage

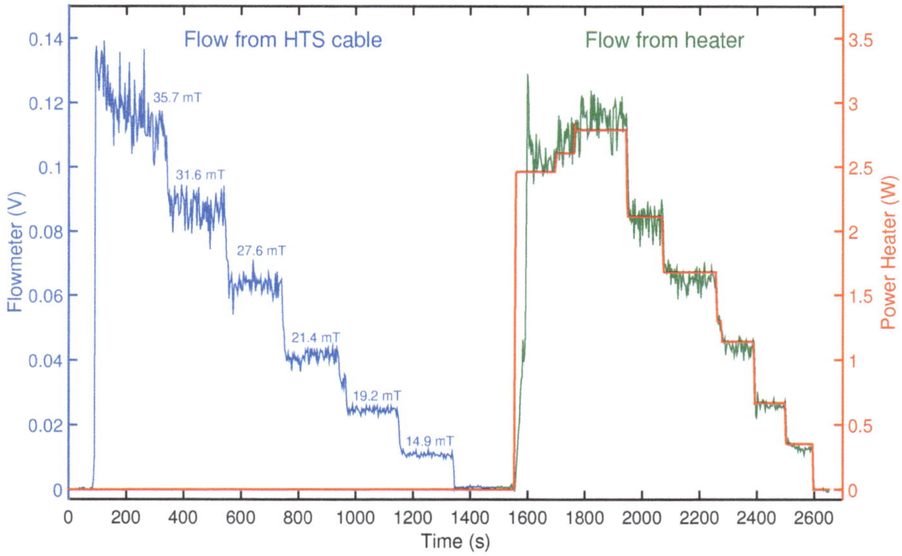

Figure 3.11: Evaporation level of the AC losses (left blue curve) for applied fields of 35.7 mT, 31.6 mT, 27.6 mT, 21.4 mT, 19.2 mT and 14.9 mT , and the calibration system (right green curve). The power of the heater according to the evaporation level is also shown (red curve).

and the current are recorded in order to calculate the power dissipated in the heater (right part of the plot in Fig. 3.11). Then for each level a coefficient is calculated, which is the result of the power from the heater divided by the voltage from the flowmeter. Finally, the power of AC magnetization losses is calculated by multiplying this latter coefficient by the flowmeter's voltage corresponding to the evaporation of the different applied magnetic fields.

4 Critical Current Measurement

The critical current is the most prominent property used to evaluate the performance of superconducting tapes and cables, and it is very important to have reliable methods to measure it. Even in the context of AC losses, a low-loss conductor needs to have a sufficiently high current-carrying capability in order to be interesting for applications. This chapter describes the experimental method to measure the critical current and the measurement results of several REBCO tapes and cables.

The critical current of REBCO tapes strongly depends on the magnitude and orientation of the magnetic field and the temperature. In this work, the focus is on power applications such as cables, fault current limiters and transformers, so that the temperature is assumed to be constant at 77 K and the magnetic field is less than 1 T. Depending on the specific application, the REBCO tape experiences a wide range of field orientation and that is why the setup is built to measure all angles.

Due to the relatively high cost of the REBCO tapes, it is important to optimize the critical current of the cables. This can only be done with a precise knowledge of the angular dependance of the critical current of the REBCO tapes. Due to the specific microstructure, REBCO tapes exhibit very anisotropic dependence of the critical current on the magnetic field. This anisotropic behavior varies considerably among different manufacturers.

Researchers have been demonstrating a reduction of the anisotropy by the so-called doping and by introducing artificial pinning centers. A general enhancement of the in-field critical current is observed in REBCO HTS tapes with additions of $BaZrO_3$ [CSZ+09, KGL+06, MM10, SCX+09, SCZ+12], $BaSnO_3$ [VBW+08], Gd_3TaO_7 [HDM+09], Y_2BaCuO_5 [HBW+04] and Ba_2YNbO_6 [FHM+10, WGZ+10]. Some manufacturers such as SuperPower [Supc] have been investigating and successfully reduced this anisotropy
[SCZ+12, SXL+14]. SuperPower's state-of-art REBCO HTS tape is commercially labeled "Advanced Pinning (AP)" [HZKS15, NZB+15, NBS+16, SXL+14, ZLF+13]. However, the angular dependence of those HTS tapes varies in a wide range and it is difficult to predict precise behavior.

For REBCO tapes without artificial pinning centers, the angular dependence can be described by the following formulas [GSZV14]:

$$I_c(B_{\parallel}, B_{\perp}) = I_c(0)\left(1 + \frac{\sqrt{B_{\perp}^2 + (kB_{\parallel})^2}}{B_c}\right)^{-b} \tag{4.1}$$

where $I_c(0)$ is the self-field critical current, $B_{\parallel}$ and $B_{\perp}$ are, respectively, the parallel and perpendicular components of the magnetic field, B_c is a sort of 'critical' field defining how fast the current degrades, k is an anisotropy parameter, and b is a damping parameter. This formula easily identifies the maximum and the minimum values of the critical current.

As mentioned before, the angular dependence of the critical current on the magnetic field of REBCO tapes changes dramatically with artificial and non-artificial pinning centers [KGL+06, PVGŠ11, KIN+02, UKSK09, ZSC+09, SCX+09, SCZ+12, SXL+14, WGZ+10, FHM+10], [HDM+09, CSZ+09]. A vast collection of data with angular dependence of the critical current can be found in the database [WS17, The16] created by N. M. Strickland and S. C. Wimbush. This database published first in September 2016, includes commercially available superconducting wires from American Superconductor [Supa], Fujikura [Fuj], Superconductor Technologies Incorporation [STI], SuNAM [SuN], SuperOx [Supb], and SuperPower (AP) [Supc]. Those data-sets can be used for modeling and designing HTS devices and are continually expanded with new data from its creators and also external contributions. Nevertheless, due to the strong variations, the characterization of state-of-art REBCO tapes is necessary on a case by case basis. That is why an in-house experimental facility has been built at the Institute for Technical Physics.

The measurement of the critical current for REBCO tapes and cables is based on the same principle. It consists in supplying the sample with a slow ramp of current while monitoring the voltage – the so-called four-point measurement. When the current approaches the critical value a measurable voltage rapidly develops. The critical current is commonly defined as the current at which an electric field of $1\,\mu V\,cm^{-1}$ is reached, as shown in Fig. 4.1. This critical current definition is standardized [Int06] for the first generation of the HTS tapes (BI-2223/Ag) and it is commonly used for the REBCO tapes as well.

In the initial transition from the superconducting to the normal state, the $I-V$ curve can be described by a power-law relation:

$$V = V_c\left(\frac{I}{I_c}\right)^{n} \tag{4.2}$$

where V_c is the critical voltage (which is $1\,\mu V\,cm^{-1}$ times the distance between the voltage taps), I_c is the critical current and n defines the steepness of the curve. The n-value indicates the gradient of the curve when the $I-V$ curve is plotted on a logarithmic scale. A high n-value represents a sharp transition from the superconducting to the normal state.

In order to measure the critical current for REBCO tapes and cables, two different experimental setups have been built: one for measuring the dependence of the critical current on the magnitude and orientation of the magnetic field; and one for measuring the self-field critical current of cables.

4.1 Critical Current Dependence of HTS Tapes

4.1.1 Experimental Setup

The critical current measurement of REBCO tapes exposed to a magnetic field is performed with a fully automated setup, shown in Fig. 4.2. The magnetic field can be varied in amplitude and in angle. The setup consists of four different parts: the generation of the magnetic field, the rotation of the sample, the ramping of the current with the recording of the voltage, and the fit of the $I-V$ curve. A LabVIEW program was written in order to make the measurements fully automatized.

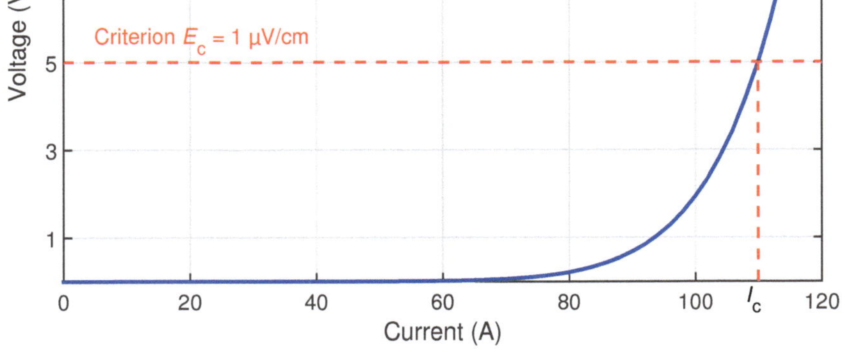

Figure 4.1: Plot of the power law using the following parameters: $I_c = 110\,\mathrm{A}$ and $n = 10$. The voltage taps were placed on 5 cm apart.

The magnetic field is generated by two copper coils with an iron core. The uniformity region where the sample is located is 10 cm. The coils are energized by a power supply (Bruker B H11 D), which can generate up to 600 mT. The coils are cooled down by a water cooler (Bruker B MN C5). The communication between the computer and the power supply of the coils is carried out by an analog-digital converter (NI-USB 6501). A Hall sensor (model LHP-NP) is placed next to the sample in order to measure the magnetic field and this sensor is connected to a voltmeter (Keithley 2000).

A double-walled cylindrical cryostat is located in the bore of a stepper motor (MiCos PRS-200), which provides the rotation. The sample holder is positioned inside it. A tray with six heating power resistors (25 W each) is placed between the cryostat and the motor in order to prevent the motor from freezing.

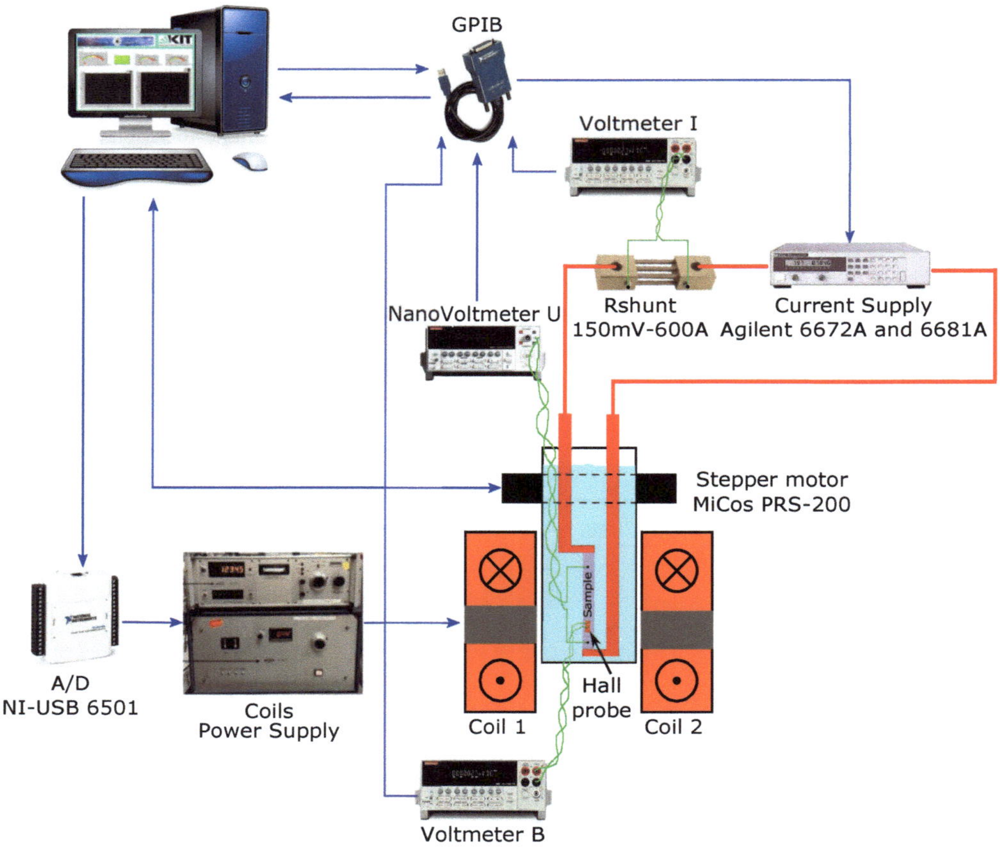

Figure 4.2: Schematic overview of the automated setup measuring the critical current as a function of the magnetic field amplitude and orientation.

The current is supplied to the sample by two current supplies (Agilent 6672A and 6680A). The Agilent 6672A is used for direct currents from 0 to 100 A and both current supplies together for direct currents higher than 100 A and up to 975 A. The current is measured with a shunt placed in series. The shunt has a ratio of 150 mV/600 A, which means that the shunt can drive a maximum current of 600 A. For that current, the voltage drop is 150 mV. The shunt is connected to a voltmeter (Keithley 2000). The voltage of the samples is measured by a nanovoltmeter (Keithley 2182A or Keysight 34420A).

A LabVIEW program was developed for controlling the hardware, setting the parameters for the measurement of the $I-V$ curve (such as time step, current step, threshold voltages), calculating the critical currents and the n-values, and collecting, monitoring and saving the data. The main feature of the program is the high flexibility for measuring the $I-V$ curve. The regular way to control the $I-V$ curve with regularly spaced current steps, as shown in Fig. 4.3(a), was not satisfying in terms of speed and accuracy. That is why two new modes of measuring the $I-V$ curve were investigated (Fig. 4.3(b) and 4.3(c)). Both modes allow defining different voltage criteria (e.g. $V_{c1} = 1\,\mu\text{V}\,\text{cm}^{-1}$ and $V_{c2} = 0.1\,\mu\text{V}\,\text{cm}^{-1}$). The n-value is then calculated between those two points.

The main idea of the first mode, shown in Fig. 4.3(b), is to simply divide the curve in two parts. In the first part the current is increased rapidly with large increments. When a measurable voltage develops, the current steps are made smaller. The second part has smaller current steps than the first one in order to get a sufficiently large number of measurement points for calculating the n-value and the critical current, and also to control the maximum injected current with more precision. The measurements are stopped when the voltage reaches 0.5-2 times V_{c1} – the particular value depending on the type of tape.

The second mode, shown in Fig. 4.3(c), divides the $I-V$ curve in three parts. The first two parts are as in the first mode described above. The third part consists in ramping the current back to zero with small current steps, and is used to calculate the n-value and the critical current. The second mode is the preferable option to use when the first part of the $I-V$ curve is affected by a large noise voltage or by a resistive slope. In those cases, if the first mode were used, the second part with fine current steps would be initiated too early at too low current values.

Once those parameters are defined and inserted in the program, the full characterization of the tape can begin. A full characterization consists in measuring the $I-V$ curves for each amplitude and orientation of the magnetic field. The main steps of the characterization process are illustrated in the flow chart of Fig. 4.4. Typically, the amplitude is varied between 25 and 600 mT and the orientation from 0° to 360° in steps of 10°. Additional angle values are often measured in order to better visualize sharp peaks in the angular dependence of I_c.

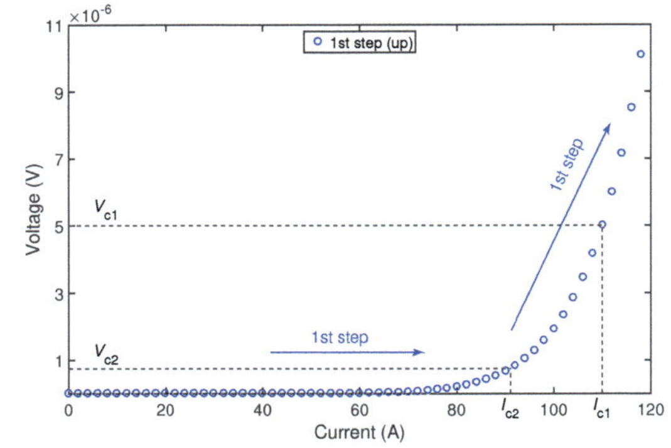

(a)

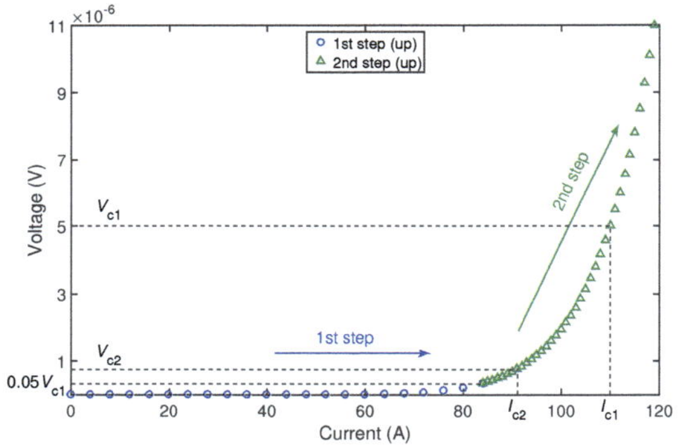

(b)

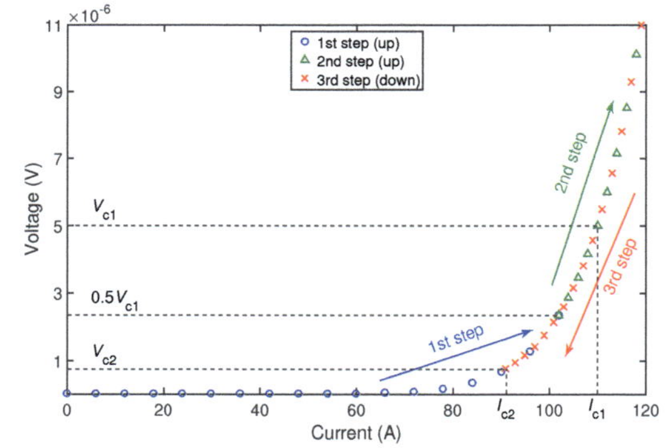

(c)

Figure 4.3: Illustration of the three different modes for plotting a $I-V$ curve. V_{c1} and V_{c2} correspond to a criterion of $1\,\mu V\,cm^{-1}$ and $0.1\,\mu V\,cm^{-1}$, respectively. The length of the voltage taps is 5 cm. (a) One-step mode (b) Two-step mode. (c) Three-step mode.

The experiment is fully automatized and allows saving a significant amount of time with respect to a manual characterization. This is because the program automatically sets each amplitude and orientation of the field. It also automatically calculates the n-values and the critical currents, as follows:

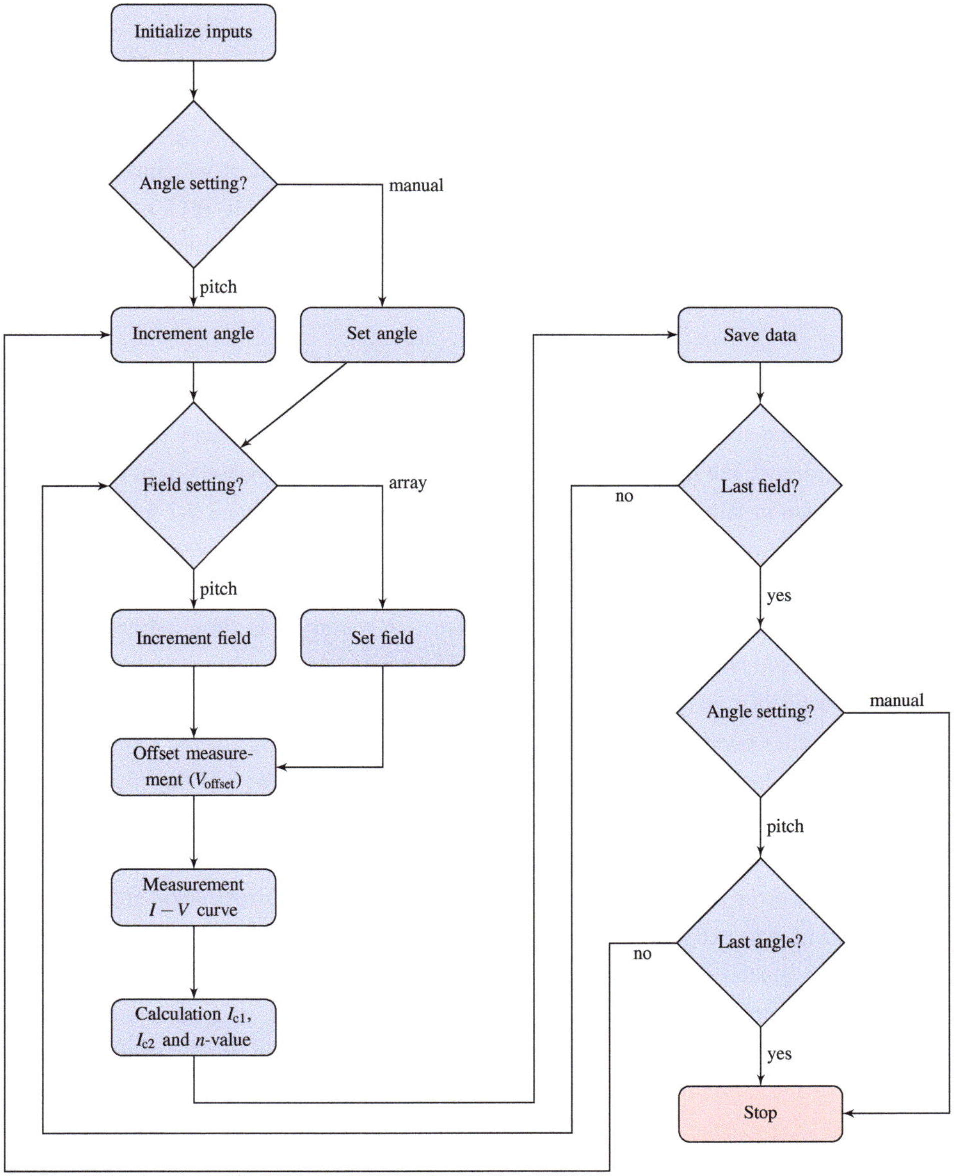

Figure 4.4: Flow chart of the angular dependence characterization of the critical current.

First, the $I-V$ data points between V_{c1} and V_{c2} are selected and the n-value is calculated with a linear fit of the slope of the curve of those points on a log-log scale. Then, for each of those selected points (I_m, V_m), a critical current value is calculated as

$$I_{c,m} = I_m \left(\frac{V_c}{V_m}\right)^{\frac{1}{n}} \tag{4.3}$$

Finally, the critical current of the sample is taken as the average of the calculated $I_{c,m}$ values.

With respect to a direct measurement of the current corresponding to the critical voltage criterion, this procedure allows a precise determination of I_c even in the case of large current increments or if the current ramp is stopped before reaching the critical voltage criterion.

This experimental facility is now routinely used for characterizing HTS tapes to be used in various HTS applications (e.g. [GSZV14, GVKZ16]) and the results serve as main input of numerical models (e.g. [ZHSG17]).

4.1.2 Measurement Results

Various input parameters for calculating the n-value and the critical currents were investigated: the current increment, the time step and the threshold voltage of the various steps of each mode. The results shown in this section refer to a current increment of 2 A and 0.2 A for a threshold voltage of $0.05V_c$ and $2V_c$, respectively for the two-step mode, and a current increment of 2 A, 0.5 A and 0.2 A for a threshold voltage of $0.5V_c$, $2V_c$ and $0.1V_c$, respectively for the three-step mode. For both modes, the time step is of 500 ms. A variation of these parameters within reasonable limits did not lead to different values of the n-value and critical current.

Different ways of applying the background field (i.e. by increasing or decreasing its amplitude) and different operating modes (i.e. by using the two-step mode or the three-step mode) were also tested. The results are shown in Table 4.1. For the n-value and the critical currents, an error margin was calculated. The error of the n-value is $\pm 4.8\,\%$, for the critical current I_{c1} is $\pm 0.6\,\%$ and for the critical current I_{c2} is $\pm 1\,\%$.

The different error margins were calculated as follow: The current error margin consists in calculating the two extreme currents from the measurement of the shunt voltage drop and its ratio, i.e. the smaller voltage with the largest ratio and the larger voltage with the smallest ratio. The shunt voltage drop was measured with a multimeter and its error is $\pm((30 * 10^{-6} * \text{Reading}) + (30 * 10^{-6} * \text{Range}_{100\text{mV}}))$. The error of the shunt ratio is $\pm 0.5\,\%$ of it. The voltage error margin of the sample consists in calculating the two extreme voltage drops of it, which was measured with a nanovoltmeter. The error of the reading of the nanovoltmeter is $\pm((20 * 10^{-6} * \text{Reading}) + (4 * 10^{-6} * \text{Range}_{10\text{mV}}))$. The n-value error margin was calculated from two extreme $I-V$ curves, i.e. firstly, when the current is the smaller and the voltage the larger, and

Table 4.1: Critical currents and n-values of the different modes for an applied field angle of 0° (⊥ to the ab-plane of tape) and a temperature of 77 K. The values correspond to the REBCO 4 mm wide tape.

a two-step mode for increasing field

B (T)	I_{c1} (A)	I_{c2} (A)	n-value
0.00	109.6 ±0.7	102.0 ±1.0	32.0 ±1.5
0.05	75.3 ±0.4	67.7 ±0.7	21.8 ±1.0
0.20	50.1 ±0.3	43.1 ±0.4	15.2 ±0.7

b three-step mode for increasing field

B (T)	I_{c1} (A)	I_{c2} (A)	n-value
0.00	109.6 ±0.7	101.3 ±1.0	29.2 ±1.4
0.05	75.6 ±0.5	66.8 ±0.7	18.7 ±0.9
0.20	50.2 ±0.3	43.3 ±0.4	15.6 ±0.7

c two-step mode for decreasing field

B (T)	I_{c1} (A)	I_{c2} (A)	n-value
0.00	109.9 ±0.7	102.8 ±1.0	34.5 ±1.7
0.05	75.4 ±0.5	67.9 ±0.7	21.8 ±1.0
0.20	50.0 ±0.3	43.4 ±0.4	16.2 ±0.8

d three-step mode for decreasing field

B (T)	I_{c1} (A)	I_{c2} (A)	n-value
0.00	109.7 ±0.7	101.2 ±1.0	28.6 ±1.4
0.05	75.6 ±0.5	67.0 ±0.7	18.9 ±0.9
0.20	50.0 ±0.3	43.4 ±0.4	16.0 ±0.8

on the other hand when the current is the larger and the voltage the smaller. The critical current error margin was calculated from the eq. (4.3) and the two latter n-values.

The results in Table 4.1 show that, within the error margin, the n-value and the critical currents are not influenced by the procedures followed for creating the $I-V$ curves (two or three-step mode) or for producing the background field (increasing or decreasing field magnitude).

Three superconducting tapes that differ in width (4 mm and 12 mm), in supplier (SuperPower and SuperOx), in technology (advanced pinning and not) and in copper stabilization (5 µm and 20 µm) were measured. The results of their full characterization are shown in Fig. 4.5.

The behavior of the samples without advanced pinning is commonly known. They have usually two main peaks and valleys at 90° and 270°, as shown in Fig. 4.5(a). The advanced pinning aims to reduce the anisotropy. The samples with this technology show a much more complicated angular dependence, as can be seen in Fig. 4.5(b) and Fig. 4.5(c).

Figure 4.5 shows that for all the considered superconducting tapes, the critical current decays with the increasing magnetic field and varies with its orientation. The angular dependence is strongly influenced by the microstructure of the superconductor. The critical currents of the SuperOx 12 mm wide tape, with 20 µm of copper-stabilization, get their greatest critical current values for the angles 90° and 270°, which correspond to the parallel orientation of the tape with respect to the applied field. In contrast, the SuperPower 12 mm wide tape, with 5 µm of copper-stabilization, gets its greatest critical currents for the perpendicular orientation (0°, 180° and 360°) of the tape with respect to the field. The third tape, the SuperPower 4 mm wide, with 5 µm of copper-stabilization, combined the two features of the two previous tapes. Indeed, this tape has its greatest critical current values for the parallel orientation (90° and 270°) with an applied field up to 300 mT and for the perpendicular orientation (0°, 180° and 360°) with an applied field from 300 mT up to 600 mT.

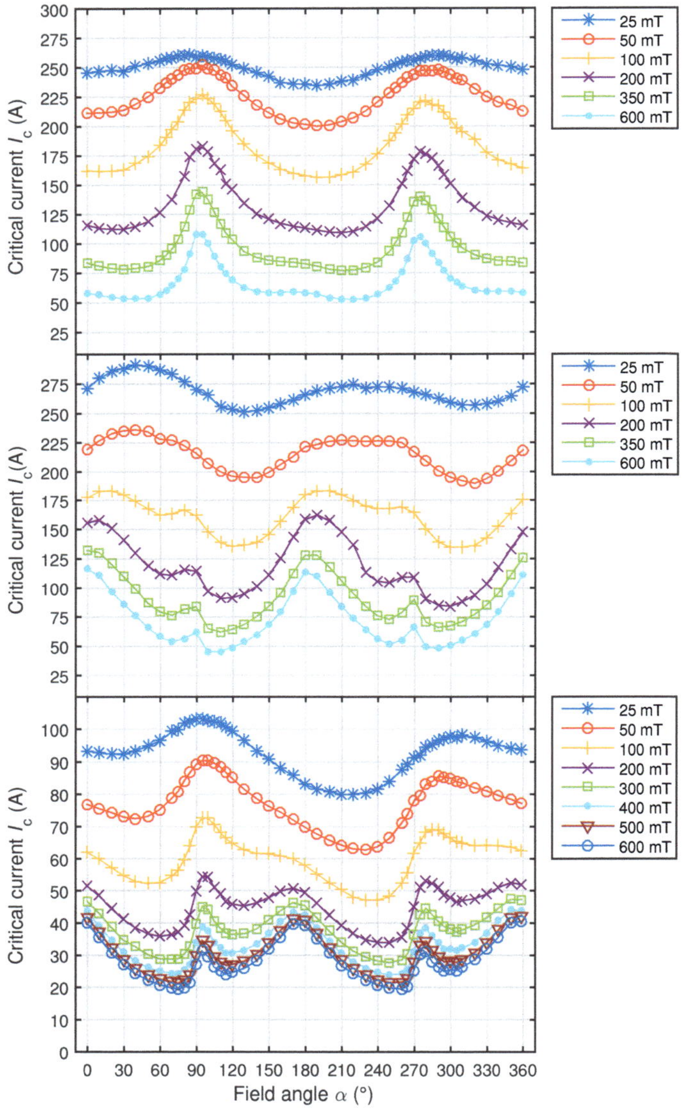

Figure 4.5: Example of angular dependence of the critical current for several applied magnetic field. (a) SuperOx 12 mm width tape (ID: SOx-J-P1-12-15Ag-20Cu-60V), (b) SuperPower 12 mm width tape (ID: SP-KIT-20101001-1c) and (c) SuperPower 4 mm width tape.

4.2 Critical Current of HTS Cables

Measuring the critical current of an HTS cable follows the same principles as for an HTS tape. Due to the presence of multiple strands, the current injection requires properly sized and designed current leads instead of a simple soldering. In addition, the voltage signal used to build the $I-V$ curve can be picked up from different locations. For these reasons, the critical current measurement of HTS cables cannot fit in the apparatus for single tapes shown in Fig. 4.2 and is performed in a dedicated larger container.

Another dedicated LabVIEW program was used for monitoring and recording the $I-V$ curves as well as stopping manually the current injected into the cable.

4.2.1 Experimental Setup

For both CORC® and RACC cables, the position of the voltage taps is illustrated in Fig. 3.7. The taps are soldered on each tape or strand and inside the current leads. For the CORC® cable the voltage taps were placed only on the outer tape layer. The different positions of the voltage taps allow monitoring the global behavior of both cables.

In order to reach sufficiently high currents, a 2 kA Heinzinger power supply (model PTN3p 6-2000) was used. The voltage signals were measured with a multiplexer (model Agilent 34970A) connected to a nanovoltmeter (model Keithley 2182A). The current flowing in the cable was monitored with a shunt resistance (ratio: 60 mV/1500 A) connected to a voltmeter (model Keithley 2000).

For this setup, a LabVIEW program is used to control all the hardware in order to monitor and save the $I-V$ curves. The program uses only one current increment setting (manually adjustable during the measurement) and the current ramp is stopped by the user, based on the behavior of the individual strands.

The use of another program with respect to the one used for the automatic I_c measurement of tapes is necessary for several reasons. Firstly, the current sharing between the strands, which influences the estimation of the critical current of the cable, is difficult to know [VGT+11, GFK+09]. That is why the transition of the $I-V$ curves along the cable can not be predicted and hence the threshold voltage for stopping the current can not be set in advance. Secondly, the current imbalance between the strands can easily cause damage in individual strands, especially if they are not sufficiently stabilized. Finally, the damage of one strand would require a complete re-assembling of the cable, with substantial loss of material and time. That is why the preferable option for measuring the $I-V$ curves of REBCO cables is the manual monitoring of all the voltage taps in order to see the behavior of each of them.

One challenge of these measurements consists in getting a sufficient number of data points after observing the rise of the voltage and shutting the current off before it reaches values that can damage the cable.

4.2.2 Measurement Results

The critical current measurements of the cables were performed in liquid nitrogen (77 K) at self field. As mentioned above, all the voltage taps are monitored as a function of the current ($I-V$ curves) and the measurement is stopped by the user in order to prevent damaging the cable. It can happen that the voltage taps do not achieve their threshold voltage. This can be due to three reasons: firstly, a strand does not start to develop a sufficient voltage yet. Secondly, the measurement is stopped before the strand reaches the threshold voltage because other strands did. Thirdly, the voltage taps have lost the soldered connection as a consequence of the cable handling, of its immersion into the liquid nitrogen bath, or both things together.

Critical Current Measurement of CORC® Cables

The distance of the CORC® voltage taps soldered on the tapes (labeled "T_{C1}", "T_{C2}" and "T_{C3}") is 144 mm. As for the voltage taps on the copper leads, labeled "L_{C0}" and "L_{C1}", their distances are arbitrarily chosen to be 200 mm and 340 mm. The exact positions of these taps are not known, apart from the fact that they are located somewhere in the inner sides and in the ends of the copper leads. This adds some uncertainty to the I_c values.

Figure 4.6(a) shows the $I-V$ curves of the five voltage taps of the CORC® cable C_{S4} (12 striated tapes on 4 layers). One can observe that the transition of the contact "L_{C0}" and "L_{C1}" starts as soon as the increase of the current and is proportional to the current. This transition, in the shape of a resistive slope, is due to the resistive joint of the current leads. The resistances for both sets of contacts are calculated and the resistive slopes are subtracted from the total voltages in order to calculate the n-value and the critical current. As for the voltages from the tapes, an offset for each pair of taps (before the transition) was calculated and subtracted from the total voltages. The resistance value is the coefficient of the linear fit of the resistive slope calculated in a range of reasonably measured points. The range changes between cable geometries (number of layers), but it is the same between the cable types with the same number of layers (reference and striated cables).

For a comparison of the critical currents from the different voltage taps, which have different lengths, the $I-V$ curves are normalized by their length, as shown in Fig. 4.6(b). This normalization already includes the subtraction of the resistive slopes for the sets of contacts located in the copper leads and of the offset for the sets of contacts located on the tapes.

As observed in Fig. 4.6(b), the superconducting transitions of the taps "L_{C0}" and "L_{C1}" are much slower than those of the taps located on the three tapes. This quick transition of the tapes compared to the global cable, indicates a higher n-value. The value equals to 1 represented by the dashed line in the y-axis corresponds to the threshold voltage (electric field criterion of $1\,\mu V\,cm^{-1}$) for all contacts. One can observe that only the contact "L_{C0}", which is also the first

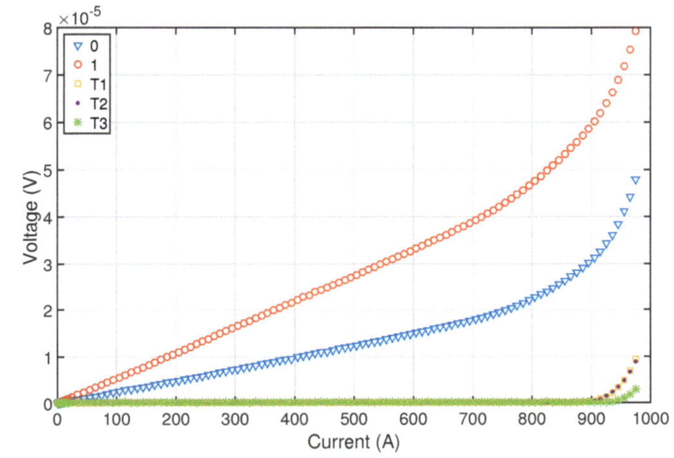

(a)

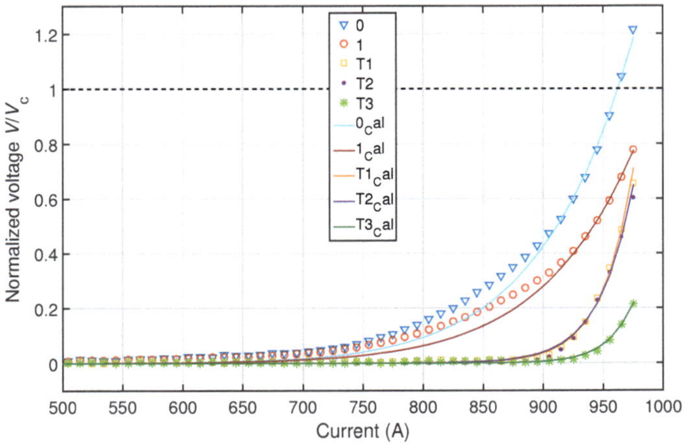

(b)

Figure 4.6: Plots of the $I-V$ curve measurement from the different voltage taps of the REBCO CORC-S4 cable, assembled with twelve wound tapes. (a) The actual measured $I-V$ curves, (b) Normalization of the measured $I-V$ curves. For the contact "L_{C0}" and "L_{C1}", the shown $I-V$ curves include the subtraction of their resistive slope. The fits of the $I-V$ curves calculated from the power law with the calculated n-values and I_c are also shown (full lines).

Table 4.2: Critical currents and n-values of each voltage taps of the prepared CORC® cables measured at 77 K. Each layer consists of three tapes. The cables labeled with "R" represent the ones with unstriated strands and those with "S", striated strands with five filaments, The number corresponds to the number of strands. The contact resistances, labeled "R_c", for the voltage taps "L_{C0}" and "L_{C1}" are also shown in (nΩ).

Cables	Contacts													
	L_{C0}			L_{C1}			T_{C1}		T_{C2}		T_{C3}			
	I_c	n	R_c	I_c	n	R_c	I_c	n	I_c	n	I_c	n		
CORC − C_{R6}	454	7.8	37.4	485	8.3	50.4	—	—	—	—	—	—		
CORC − C_{S6}	410	15.8	27.2	419	12.6	32.2	432	17.2	—	—	—	—		
CORC − C_{R9}	795	9.4	114.5	805	6.1	126.3	805	28.7	—	—	824	21.7		
CORC − C_{S9}	588	10.3	37.1	619	11.6	65.6	—	—	—	—	—	—		
CORC − C_{R12}	998	12.4	39.3	1023	11.7	107.8	1045	35.2	—	—	1032	31.9		
CORC − C_{S12}	963	13.1	24.2	995	12.9	54.1	984	37.3	987	34.9	1008	44.7		

ID: SP-KIT-20140429-2b

one, crosses the threshold voltage, because the current was switched off before the others cross this limit.

Due to the two different transitions (from the current leads and from the tapes), two ranges of criteria are used for calculating the n-value of each $I-V$ curve. The range used for the taps "L_{C0}" and "L_{C1}" is between $0.5\,\mu V\,cm^{-1}$ and $1.5\,\mu V\,cm^{-1}$. On the other hand, the contacts of tapes use a range between $0.1\,\mu V\,cm^{-1}$ and $1\,\mu V\,cm^{-1}$. The critical current of each point located between the two criteria of each $I-V$ curve is calculated from the eq. (4.3) using the n-value calculated previously. The total critical current of a cable is then considered as the mean value of those calculated values. In order to see the accuracy of the parameters calculated, new $I-V$ curves are plotted using eq. (4.2) in full line in Fig. 4.6(b).

All the calculated values such as the critical currents, the n-values and the contact resistances are summarized in Table 4.2.

Critical Current Measurement of RACC Cables

The distance of the RACC voltage taps soldered on the strands (labeled from "T_{R1}" to "T_{R10}") is 252 mm. As for the voltage taps on the copper leads, labeled "L", their distances are arbitrarily chosen to 378 mm. Similarly to the case of the CORC® cable, the exact positions of these taps are not known and this adds some uncertainty to the I_c values.

Figure 4.7(a) shows the $I-V$ curves of the eleven voltage taps of RACC-SP-S5 cable (ten striated strands). One can observe that the transition of the contact "L_R" starts as soon as the increase of the current and it is proportional to the current. This transition, as also observed in CORC® cables, is due to the resistive joint of the current leads. The resistance for this set of contacts is calculated and the resistive slope is subtracted from the total voltage in order to calculate the n-value and the critical current. As for the voltages from the strands, an offset for each pair of taps (before the transition) was calculated and subtracted from the total voltages.

Table 4.3: Critical currents and the n-values of the prepared non-striated and striated RACC cables measured at 77 K. The cables consist of ten strands. The contact resistance, labeled "R_c", for the voltage taps "L_R" are also shown in (nΩ).

Contacts		RACC cables	
		Non-striated (RACC − R_{R10})*	Striated (RACC − R_{S10})*
L_R	I_c	1004	1040
	n	4.7	20.6
	R_c	12.6	15.6
T_{R1}	I_c	1199	1020
	n	47.0	21.0
T_{R2}	I_c	1201	1016
	n	48.0	21.0
T_{R3}	I_c	1196	993
	n	42.6	12.2
T_{R4}	I_c	1176	1027
	n	23.7	25.0
T_{R5}	I_c	1098	1014
	n	7.8	19.6
T_{R6}	I_c	1074	1036
	n	6.5	26.9
T_{R7}	I_c	914	1032
	n	5.3	29.7
T_{R8}	I_c	1181	1016
	n	23.9	19.1
T_{R9}	I_c	1199	1013
	n	48.4	18.6
T_{R10}	I_c	/	/
	n	/	/

*ID: SOx-2015-43

The resistance value is the coefficient of the linear fit of the resistive slope calculated in a range of reasonably measured points.

For a comparison of the critical currents from the different voltage taps, which are different lengths, the $I-V$ curves are normalized by their length, as shown in Fig. 4.7(b). This normalization already includes the subtraction of the resistive slopes for the sets of contacts located in the copper leads and of the offset for the sets of contacts located on the strands. The value equals to 1 represented by the dashed line in the y-axis of the Fig. 4.7(b) corresponds to the threshold voltage (electric field criterion of 1 μV cm^{-1}) for all contacts.

All $I-V$ curve transitions have the same behavior, that is why with respect to the CORC® cables, one voltage range is used for calculating the n-value of each $I-V$ curve. The used range

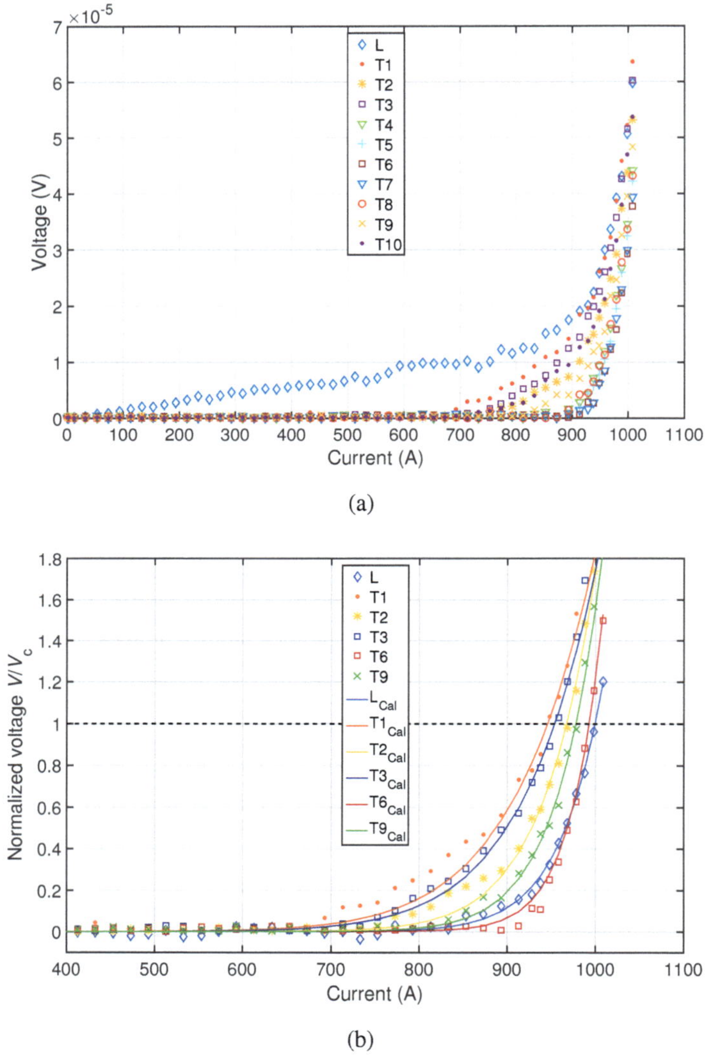

Figure 4.7: Plots of the $I-V$ curve measurement from the different voltage taps of the REBCO RACC-SP-S5 cable, assembled with ten meander strands. (a) The actual measured $I-V$ curves, (b) Normalized $I-V$ curves of a few voltage taps. The $I-V$ curve of the contact "L_R" includes the subtraction of its resistive slope. The fits of the $I-V$ curves calculated from the power law with the calculated n-values and I_c are also shown (full lines).

Table 4.4: Sum of the I_c from the single strands used for assembling the cables that is compared to the I_c measured of their cable. The self field degradation of the critical current is also shown. The first six rows correspond to the CORC® cables with the measured critical current from the contact "L_{C0}". The two last rows correspond to the results of RACC cables, the measured critical current is from one of the strands.

Cables	Sum I_c tapes (A)	Standard deviation	I_c measured (A)	Self field reduction (%)
$CORC-C_{R6}$	621	1.5	454	26.9
$CORC-C_{S6}$	589	3.6	410	30.4
$CORC-C_{R9}$	935	1.9	795	15
$CORC-C_{S9}$	894	5.3	588	34.2
$CORC-C_{R12}$	1243	1.7	998	19.7
$CORC-C_{S12}$	1206	1.8	963	20.1
$RACC-R_{R10}$	1461	2.7	914	37
$RACC-R_{S10}$	1298	5.6	993	24

is between $0.5\,\mu V\,cm^{-1}$ and $1.5\,\mu V\,cm^{-1}$. The critical current of each point located between these two voltage values of each $I-V$ curve is calculated from eq. (4.3) using the previously calculated n-value. The total critical current of a cable is then considered as the mean value of those latter calculated values. In order to see the accuracy of the parameters calculated, new $I-V$ curves are plotted using eq. (4.2) in full line in Fig. 4.7(b).

All the calculated values such as the critical currents, the n-values and the contact resistances of the RACC cables are summarized in Table 4.3. From this table, one can observe that the smallest critical current of each cable is observed from one of its strands. The n-value corresponding to this latter is also the smallest value.

In order to avoid a damage of the cables, the smallest calculated critical current is taken into account. For CORC® cables, this critical current is the value of the contact "L_{C0}" and for RACC cables the value of the contact of one of the strands. These measurements show that the critical current of a CORC® cable can be measured with one single voltage tap placed on the inner side of the current leads and not with the ones located on the tapes as also observed in [vdLNM+13]. But regarding a RACC cable, the measurement of all voltage taps can not be avoided because one can not predict which strand will give the smallest critical current.

Critical Current Analysis

The critical currents of all tapes and strands composing the cables were individually measured. For each cable, the sum of the critical currents of the composing tapes or strands was calculated. For each cable the sum of its single tapes or strands were calculated. This sum was compared to the critical current of its corresponding cable previously measured. A reduction value was defined, which corresponds to the difference (in percent) between the sum of critical currents of the single strands and the critical current of the cable. This reduction value is called "self field reduction". A standard deviation value was also calculated for all the cables, which indicates the amount of variation of critical currents of the strands, in order to evaluate the uniformity of the composing strands and its influence on the self-field reduction. These latter values are summarized in Table 4.4.

For a given number of layers of CORC® cables, the self field reduction value of the striated CORC® cable is higher than that of the non-striated CORC® cables. The standard deviation value, for a given number of layers, increases for cables made with striated cables. This explains why the self field reduction value is higher with striated cables. On the other hand, the self field reduction value of RACC cables decreases with the striated cable despite the fact that its standard deviation values is higher than that of the non-striated cable. The reason of this is unknown.

5 AC Magnetization Loss Analysis of HTS Striated Coated Conductors

In this chapter the applicability of the laser ablation technique for producing filaments on copper-stabilized coated conductors is investigated. In order to produce the filaments, two processes were used: Striated After Electroplating (SAE) and Striated Before Electroplating (SBE). These processes were applied to coated conductors with two thicknesses of copper stabilization, 5 µm and 10 µm thick. More details about the preparation of the samples can be found in section 3.1.

From the two processes and the two thicknesses of copper, four batches of coated conductors were prepared. A batch consists of one unstriated tape and four striated tapes from 10 filaments to 60 filaments, using one of the two striation processes and one of the two thicknesses of copper. For this investigation, the samples are from one manufacturer only, in order to facilitate the comparison between preparation methods. Tapes with 12 mm width were chosen for the possibility of producing a certain variety of numbers of filaments (from 10 to 60).

In section 5.1 the quality of the groove and the critical current of the tapes are shown. The groove quality of the 10-filament tape of each batch is analyzed. Cross section pictures of several grooves, one per batch are made for this analysis. This allows measuring the thickness of each material layer surrounding the groove and helps to interpret and support the AC loss measurements. The critical current of each sample with respect to the number of the filaments is plotted so that the influence of the striation on the critical current is analyzed. The values of the copper thicknesses and the critical currents are used in the subsequent investigations.

In section 5.2 the normalized AC magnetization losses are investigated. The influence of the number of filaments, the frequency and the striation process on the AC magnetization loss are shown. First the AC magnetization losses of all batches are measured as a function of the applied field at a defined frequency. These plots allow observing the influence of the number of filaments. Then two batches are taken for a more detailed look at the influence of the frequency. For the two batches, the AC magnetization losses as a function of the applied field are measured at four frequencies. Those batches differ in the striation process, but they have a similar copper stabilization thickness. Finally two striated tapes (10 filaments and 40 filaments) from each batch are taken for a more detailed look at the influence of the striation process. The AC magnetization losses of those two tapes are plotted as a function of the applied field at two different frequencies (6 Hz and 130 Hz) and compared to an analytic model that represents the lowest limit of the hysteresis losses for uncoupled filaments.

Finally, in section 5.3 the contribution of the coupling current to the AC losses is investigated in order to interpret and support the AC magnetization loss measurements. The transverse resistivity between the filaments is estimated from the AC loss measurements and compared to that directly measured in DC experiments. The resistivity is calculated for each 10-filament tape.

At the end of the chapter a summary of these investigations is presented.

5.1 Tape Properties

5.1.1 Quality of the Groove

Figure 5.1 shows Scanning Electron Microscope (SEM) pictures of the Focused Ion Beam groove milling of the SAE and SBE 10-filament tapes with different copper thicknesses. The principle of the Focused Ion Beam cut is detailed in [RP01]. The two layers of platinum visible in all the pictures were added before Focused Ion Beam cut in order to distinguish the ablation materials of the filament laser cutting from those resulting from Focused Ion Beam cutting. In the case of both SBE samples, oxygenation was applied before copper plating, but it did not result in copper-free laser grooves.

The actual thicknesses of the layers such as superconductor, silver and copper of the four 10-filament samples were measured from Focused Ion Beam cuts. The values are summarized in Table 5.1. The actual thicknesses of the copper stabilization are different from the expected values. This is due to a technical difficulty with the electroplating technique: namely, the sensitive control of the duration of a voltage pulse applied to the cell where the aqueous solution is located [FDRW$^+$11]. Other parameters of the process, such as the concentration of the sulfuric acid and of the copper sulfate solution, can also critically influence the final copper thickness [FDRW$^+$11]. Throughout this chapter the theoretical values of 5 µm and 10 µm are kept for labeling the samples.

In the SAE-5 µm tape (Fig. 5.1(a)), the groove is homogeneous in width (20 µm) and it is structured by laser down to the Hastelloy layer. There are defects visible on both sides of the groove (pointed out by arrows in the picture), but their composition could not be identified. For the determination of the groove's width, these defects are treated as non-existing material.

In the SBE-5 µm tape (Fig. 5.1(b)), the superconducting filaments are still about 20 µm apart, but the groove is partly covered by the copper.

In the SAE-10 µm tape (Fig. 5.1(c)), the groove is not well defined: along its length (direction perpendicular to the Focused Ion Beam cut), one can observe an uneven surface at the bottom of the groove, composed of melted materials from the original layers. These materials help the current to flow between filaments and are therefore cause of electrical coupling.

In the SBE-10 µm tape (Fig. 5.1(d)), the copper is completely deposited on the groove. In this case, on the left and right side of the groove, cracks between Ag and REBCO layer are observed and pointed out by arrows in the picture. The influence of those defects on the superconductor's quality cannot be identified with analysis methods presently available. Those cracks were not excluded from the definition of the laser groove width.

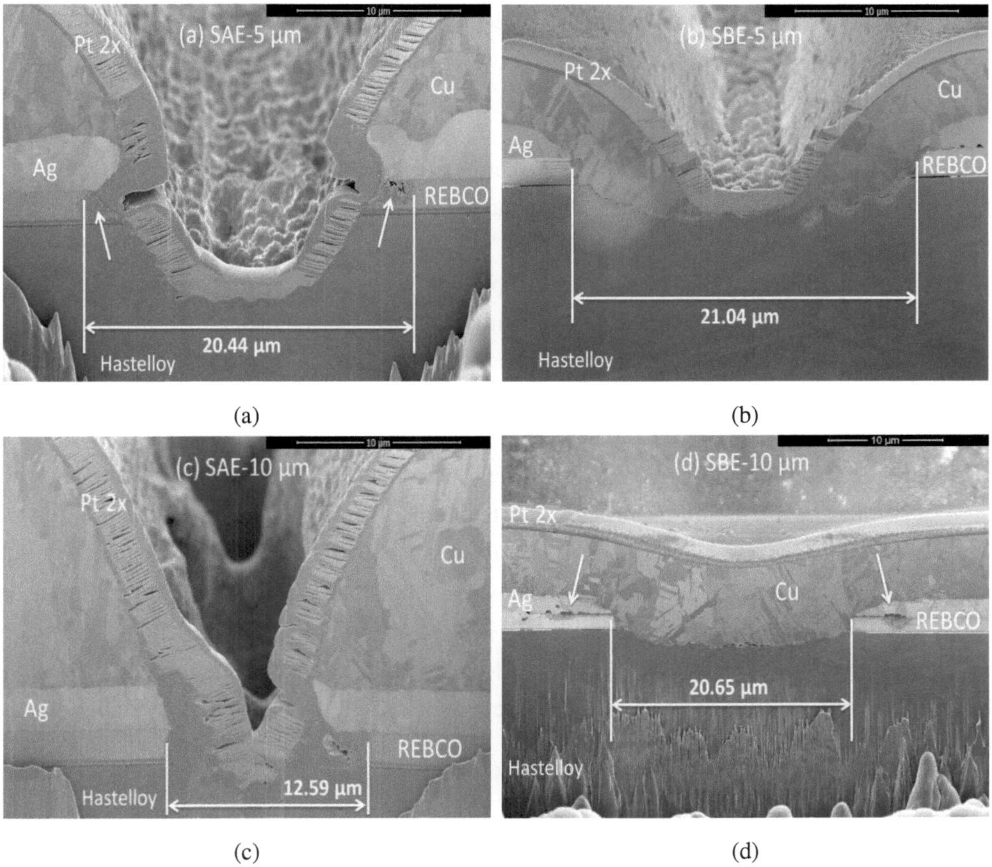

Figure 5.1: Cross section of one laser groove from 10-filament tapes. The Scanning Electron Microscope (SEM) pictures were taken from one of the nine grooves of the tapes. The thickness of the copper can vary along the tape. The values indicated here are the expected values and are used to identify the samples. (a) SAE-5 µm, (b) SBE-5 µm, (c) SAE-10 µm, (d) SBE-10 µm.

Table 5.1: Thicknesses of copper (t_{Cu}), silver (t_{Ag}) and superconductor (t_{SC}) layers in the samples shown in Fig. 5.1.

	SAE-5 µm	SBE-5 µm	SAE-10 µm	SBE-10 µm
t_{Cu} (µm)	8.6	3.5	15.9	8.1
t_{Ag} (µm)	2.6	1.7	2.3	1.8
t_{SC} (µm)	1.5	1.2	1.5	1.3

From the SEM pictures, it is clearly visible that the SAE process with thicker copper layer (SAE-10 µm) leads to a redeposition of the material in the groove and it is geometrically inhomogeneous. In the SBE-5 µm case, electrodeposited copper goes partly into laser groove. In the SBE-10 µm case, copper is deposited on the entire surface of the sample.

5.1.2 Critical Current

In general, filamentized tapes have lower I_c than the original unstriated sample [PAO16], and this has multiple causes: the loss of superconducting material, the inhomogeneity of the superconducting material [AMM+06, GBM07, SPŠ+13] and the possible damage caused by laser ablation. In the case of SBE samples, electroplating of the copper can also lower I_c: as explained in [FDRW+11], the copper electroplating technique uses an aqueous $CuSO_4$ solution with sulfuric acid H_2SO_4. If the REBCO surface is exposed to this acid, it can get damaged. In the case of SBE samples, making grooves opens the REBCO layer, which becomes vulnerable to an exposure to the acid.

Figure 5.2 shows the measured critical currents as a function of the number of filaments for the four batches. The continuous line represents the theoretical critical current that the SAE-5 µm tapes would have, based solely on the loss of superconducting material caused by the laser ablation. Regarding the other kinds of tapes, the theoretical lines have similar slope but are shifted down or up to their reference tape. For sake of readability, they are not displayed.

Figure 5.2 indicates that there is not a well-defined dependence of the critical current on the striation process. When thickness is considered, the sample with 5 µm has a smaller degradation (with the exception of one data point).

5.2 AC Magnetization Loss Measurement

AC magnetization losses were measured with the calibration-free method at 77 K. More detail of the measurement method can be found in section 3.1.

At sufficiently large applied fields, the hysteresis loss in the superconductor is expected to reduce proportionally to the number of filaments. Mawatari provided an analytic formula for the loss of an infinite array of superconducting strips subjected to perpendicular field, a geometry that approximates well a striated coated conductor with electrically uncoupled filaments (introducing an infinite resistance between filaments) [Maw96]. The formula was experimentally verified in [MZX+09].

Mawatari's formula represents a lower limit for the hysteresis losses of a striated tape without taking into account the coupling loss between superconducting filaments. The formula was used as a reference to evaluate the efficiency of the striation process in reducing the losses. The well-known formula by Brandt was used for unstriated tapes [BI93]. Brandt's expression can be considered as the upper limit of the magnetization AC losses for this type of tape.

When considering filamentized tapes, the AC loss reduction could partly be caused by the reduction of the critical current [AKY+04] and not by the fact of having well-separated narrow filaments. For this reason, the AC losses presented here are normalized by the critical current of the corresponding sample as shown in Fig 5.2:

$$Q_{\mathrm{Norm}} = \frac{Q_{\mathrm{Sample}}}{I_{\mathrm{c,Sample}}}. \tag{5.1}$$

In this section the normalized AC magnetization losses as a function of the magnetic field amplitude are presented. For this purpose, a reduction factor α, defined as the ratio between the normalized losses of the reference tape and those of each striated sample, is introduced:

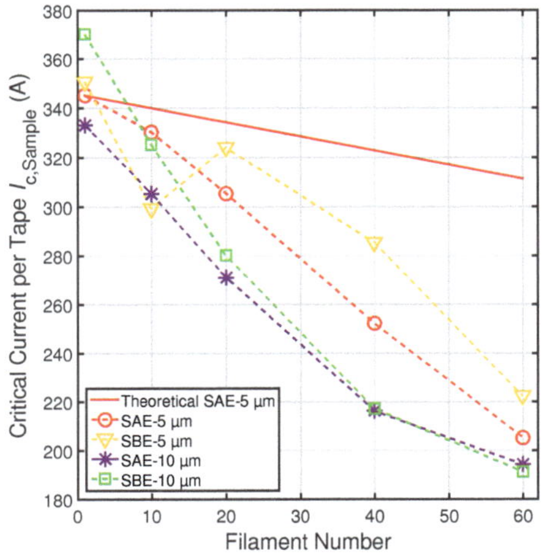

Figure 5.2: Critical currents of the different prepared tapes. The theoretical line for the SAE-5 μm tapes corresponding to the sole loss of superconducting material is also shown.

$$\alpha = \frac{Q_{\text{Norm,Ref}}}{Q_{\text{Norm,Str}}}. \tag{5.2}$$

Associated to each AC loss plot is a table listing the factor α for an applied field of 50 mT.

First the AC losses of all batches are compared at 24 Hz. This comparison allows observing the influence of the number of filaments.

Afterwards, the AC losses of two batches, SAE-5 µm and SBE-10 µm are compared at various frequencies (from 6 Hz to 130 Hz). Those two batches have different striation processes, but a very similar copper thickness, 8.6 µm and 8.1 µm, respectively. This comparison allows observing the influence of the frequency.

Then a more detailed look at the AC losses of the samples with 10 filaments and 40 filaments at two extreme frequencies, 6 Hz and 130 Hz, is given. This comparison allows observing the influence of the striation process.

5.2.1 Influence of the Number of Filaments

In Fig. 5.3, the AC magnetization losses versus the amplitude of the applied magnetic field for all the tapes are shown at 24 Hz. Table 5.2 lists the AC loss reduction factors for the different tapes for an applied field of 50 mT.

Fig. 5.3 shows that, for all the considered batches, the losses of the reference tapes are higher than the predicted ones at fields lower than 10 mT. This behavior is commonly reported in the literature [AMM$^+$06], [GBM07], [SPŠ$^+$13] and usually put in relation with the non-uniformity of the sample's properties at the edges.

The influence of the number of filaments, however, is different for the various batches:

- **SAE-**5 µm **tapes at** 24 Hz – As shown in Fig. 5.3(a), the loss reduction is higher with higher number of filaments (up to 40 filaments). The AC loss reduction of the 60-filament tape is less effective than that of the 40-filament tape, but more effective than that of the 20-filament tape.

- **SBE-**5 µm **tapes at** 24 Hz – Figure 5.3(b) shows a similar trend as the previous batch (SAE-5 µm), i.e. that the loss reduction is higher with higher number of filaments, but only up to 40 filaments. One can also observe that the AC losses of the 10-filament and 20-filament tapes are higher than those of the SAE-5 µm batch. Regarding the AC loss levels of the 40-filament and 60-filament tapes, they are very similar, as can also be seen from the AC loss reduction factors listed in Table 5.2.

- **SAE-**10 µm **tapes at** 24 Hz – In Fig. 5.3(c), one can observe two patterns for the AC losses: one for the 10-, 20- and 60-filament tapes and one for the 40-filament tape. The

AC losses of the latter are lower than those of the first pattern. The AC losses of the 10-filament tape are between those of the two previous batches (SAE-5 µm and SBE-5 µm) and the AC losses of the 20-filament, 40-filament and 60-filament tapes are higher than those of the two previous batches.

- **SBE-**10 µm **tapes at** 24 Hz – Regarding this batch shown in Fig. 5.3(d), one can also observe two AC loss patterns, but in this case the first pattern consists of the 10-filament, 20-filament and 40-filament tapes and the second of the 60-filament tape, which has AC

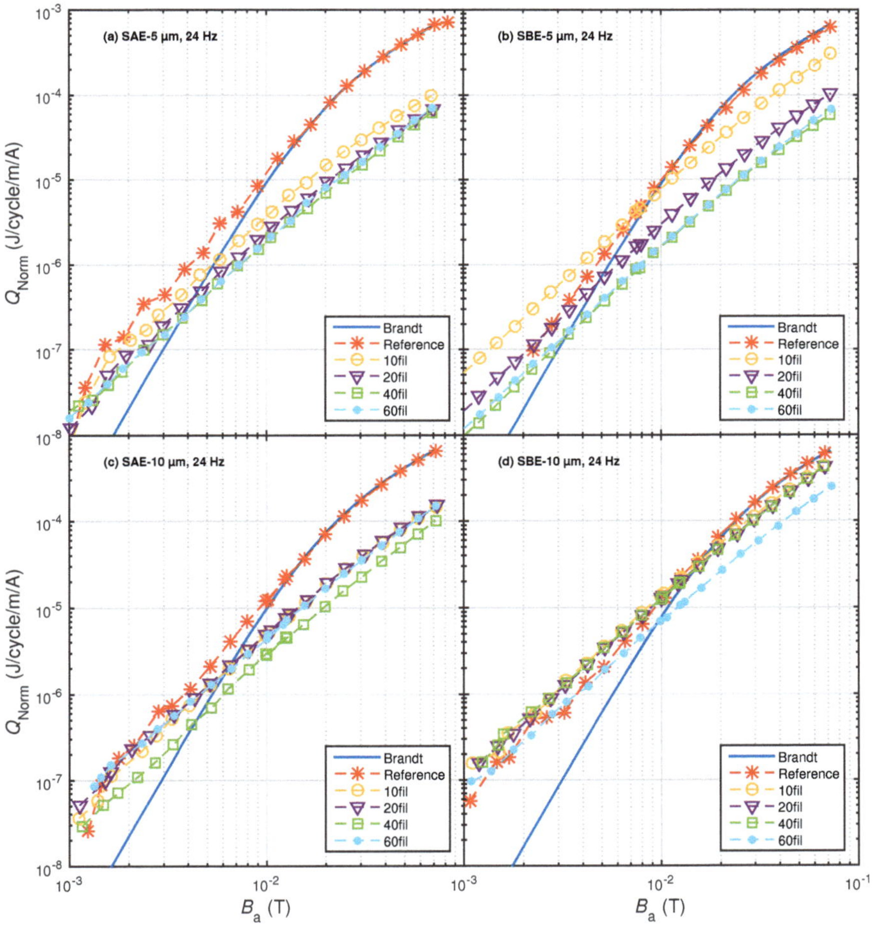

Figure 5.3: AC magnetization loss of the reference, 10-filament, 20-filament, 40-filament and 60-filament tapes at 24 Hz: (a) SAE-5 µm, (b) SBE-5 µm, (c) SAE-10 µm and (d) SBE-10 µm.

Table 5.2: AC loss reduction factor with respect to unstriated tape at 24 Hz and at 50 mT.

Process	Copper Thickness*	Number of Filaments			
		10	20	40	60
SAE	5 µm (8.6 µm)	6.9	10.2	12	11.1
	10 µm (15.9 µm)	4.8	4.5	7.6	5.1
SBE	5 µm (3.5 µm)	2.2	6.3	11.2	10.2
	10 µm (8.1 µm)	1.5	1.5	1.6	2.6

*First value is the expected copper thickness and the value in parenthesis is the measured thickness.

losses lower than those of the first pattern. The AC losses of all the tapes are higher than those of the three previous batches (SAE-5 µm, SBE-5 µm and SAE-10 µm).

The behavior of the different batches can also be seen by looking at the AC loss reduction factors at 50 mT listed in Table 5.2. The AC loss reduction is higher with higher number of filaments, but up to 40 filaments for the SAE-5 µm, SBE-5 µm and SAE-10 µm batches. For a given process (SAE or SBE), depositing a thinner copper layer allows obtaining larger AC loss reductions. The lower losses at 24 Hz are obtained with the 40-filament tapes for the SAE-5 µm, SAE-10 µm and SBE-5 µm samples, as observed in Table 5.2. The utilized process has an important influence. For example, the SBE-10 µm sample has a similar copper thickness to that of the SAE-5 µm sample (8.1 µm versus 8.6 µm), but significantly lower loss reduction, for all numbers of filaments. This is due to the large coupling losses caused by the presence of a copper layer all over the sample (Fig. 5.1(d)). A more detailed analysis of the influence of the striation process on the losses is presented in section 5.2.3.

5.2.2 Influence of the Frequency

Two batches were selected for the investigation of the influence of the frequency on the AC magnetization losses. The batches SAE-5 µm and SBE-10 µm are shown, due to the fact that they have two different striation processes but a similar copper thickness, 8.6 µm and 8.1 µm, respectively.

Figure 5.4 shows the results of the normalized AC magnetization losses with respect to the applied field amplitude regarding all samples of the SAE-5 µm batch. This figure compares four different frequencies (from 6 Hz to 130 Hz). Table 5.3 lists the corresponding AC loss reduction factors at 50 mT.

- **SAE-5 µm tapes at** 6 Hz – In Fig. 5.4(a), one can observe a reduction of the AC losses due to the striation. The loss reduction is higher with a higher number of filaments.

- **SAE-5 µm tapes at 24 Hz** – In Fig. 5.4(b), one can still observe a reduction of the AC loss level due to the striation, but the reduction of the AC loss increases with the number of filaments up to 40 filaments. The level of the AC loss of the 60-filament tape is higher than that of the 40-filament tape. The gap between the curves with the largest (40-filament tape) and smallest (10-filament tape) reduction has shrunk in comparison to the gap in Fig. 5.4(a) (6 Hz). The AC loss level of tapes is higher than that at 6 Hz, which is also listed in Table 5.3.

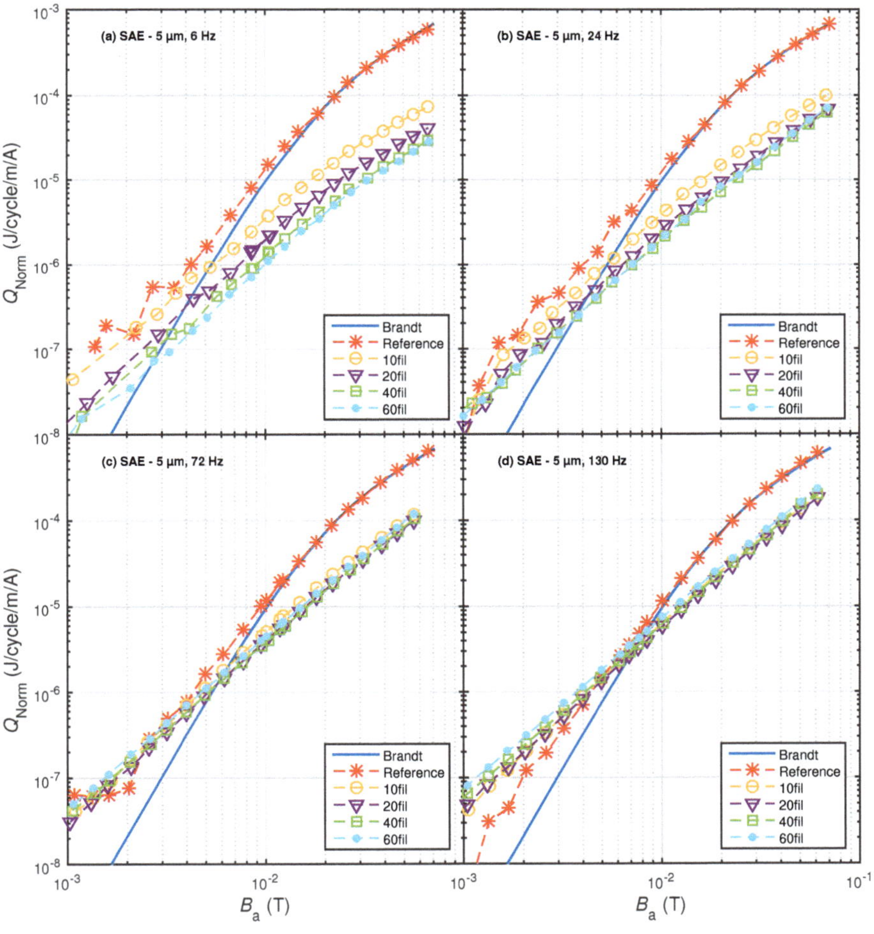

Figure 5.4: AC magnetization loss of the reference, 10-filament, 20-filament, 40-filament and 60-filament tapes of the SAE-5 µm at various frequencies: (a) 6 Hz, (b) 24 Hz, (c) 72 Hz and (d) 130 Hz.

Table 5.3: AC loss reduction factor with respect to unstriated tape for the batch SAE-5 µm for various frequencies and at 50 mT.

Process	Copper Thickness*	Frequency	Number of Filaments			
			10	20	40	60
SAE	5 µm (8.6 µm)	6 Hz	7.9	13.9	20.6	22.2
		24 Hz	6.9	10.2	12.1	11.1
		72 Hz	4.5	5.4	5.1	4.5
		130 Hz	3.4	3.6	3.0	2.8

*First value is the expected copper thickness and the value in parenthesis is the measured thickness.

- **SAE-**5 µm **tapes at** 72 Hz – In Fig. 5.4(c), one can observe a reduction of the AC loss level due to the striation, but the AC loss level of all striated tapes is similar. The AC loss level of the tapes is higher than that in Fig. 5.4(b) (24 Hz), which is also listed in Table 5.3.

- **SAE-**5 µm **tapes at** 130 Hz – In Fig. 5.4(d), one can observe a reduction of the AC loss level due to the striation but the AC loss level of all striated tapes is similar. The AC loss level of the tapes is higher than that in Fig. 5.4(c) (72 Hz), which is also listed in Table 5.3.

At 6 Hz, a lowering of the AC loss level with respect to the number of the filaments is observed. With increasing frequency the AC losses increase and the influence of the number of filaments becomes less significant, and finally disappears at 72 Hz and 130 Hz. This behavior proves the presence of filament coupling.

Figure 5.5 shows the results of the normalized AC magnetization losses with respect to the applied field amplitude regarding all samples of the SBE-10 µm batch. This figure compares four different frequencies (from 6 Hz to 130 Hz). Table 5.4 lists the corresponding AC loss reduction factors at 50 mT.

- **SBE-**10 µm **tapes at** 6 Hz – In Fig. 5.5(a), one can observe a reduction of the AC losses due to the striation. The loss reduction is higher with a higher number of filaments.

- **SBE-**10 µm **tapes at** 24 Hz – In Fig. 5.5(b), one can still observe a reduction of the AC loss level due to the striation. But two different AC loss levels can be distinguished: a first AC loss level for the striated tapes from 10 filaments to 40 filaments and a second level for the 60-filament tape. The AC loss level of the tapes is higher than that in Fig. 5.5(a) (6 Hz), which is also listed in Table 5.4.

- **SBE-**10 µm **tapes at** 72 Hz – In Fig. 5.5(c), one can observe that the reduction of the AC loss level occurs only for the striated tape with 60 filaments. The other tapes have AC losses similar to those of the reference tape. The AC loss level of the 60-filament tape is higher than that in Fig. 5.5(b) (24 Hz), which is also listed in Table 5.4.
- **SBE-**10 µm **tapes at** 130 Hz – In Fig. 5.5(d), one can observe that all tapes (the reference and the striated tapes) have the same level of AC losses. These AC losses are higher than the AC losses of the 60-filament tape in Fig. 5.5(c) (72 Hz), which is also listed in Table 5.4.

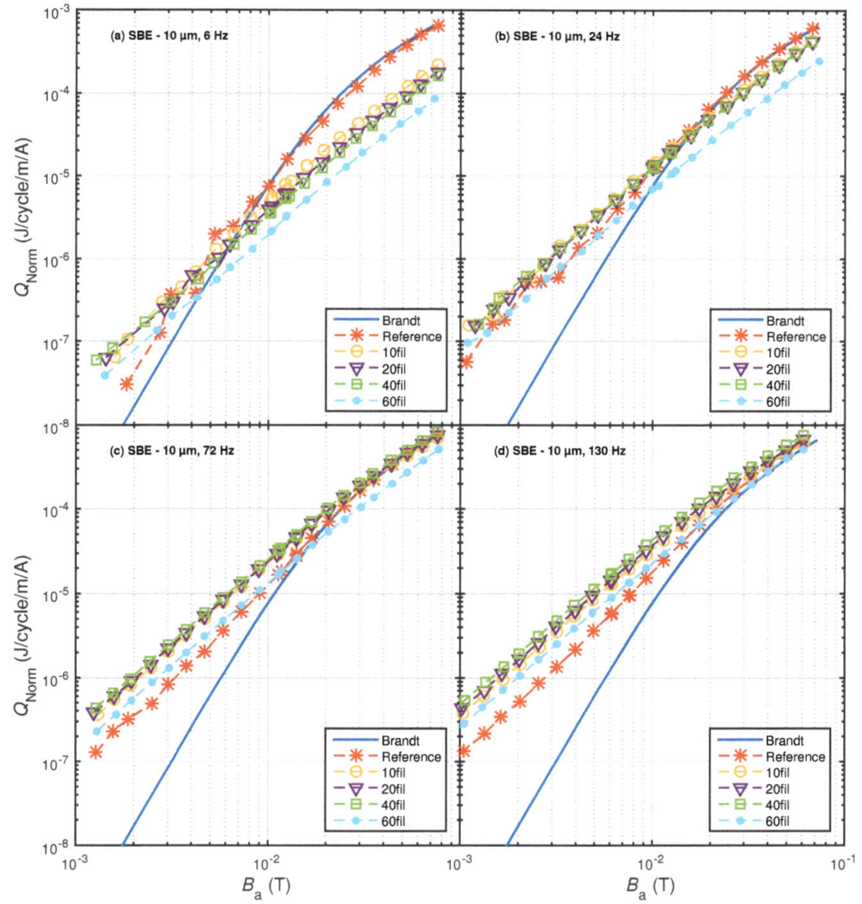

Figure 5.5: AC magnetization loss of the reference, 10-filament, 20-filament, 40-filament and 60-filament tapes of the SBE-10 µm at various frequencies: (a) 6 Hz, (b) 24 Hz, (c) 72 Hz and (d) 130 Hz.

Table 5.4: AC loss reduction factor with respect to unstriated tape for the batch SBE-10 µm for various frequencies and at 50 mT.

Process	Copper Thickness*	Frequency	Number of Filaments			
			10	20	40	60
SBE	10 µm (8.1 µm)	6 Hz	3.2	4.1	4.5	9.0
		24 Hz	1.5	1.6	1.6	2.7
		72 Hz	1.1	1.0	0.9	1.6
		130 Hz	1.0	0.9	0.8	1.1

*First value is the expected copper thickness and the value in parenthesis is the measured thickness.

At 6 Hz, a lowering of the AC loss level with respect to the number of filaments is observed. By increasing the frequency, the AC loss level increases. Two AC loss level patterns can be distinguished at 24 Hz and 72 Hz. The AC loss level of striated tapes from 10 filaments to 40 filaments are close to each other and are higher than those of the 60-filament tape. At 130 Hz all AC losses have the same level. For 6 Hz and 24 Hz, making filaments reduce the AC magnetization losses, but at 72 Hz only the 60-filament tape reduces those losses. At 130 Hz the striation does not influence the reduction of the AC losses. This behavior suggests the presence of a coupling current that is more significant for the striated tapes from 10 filaments to 40 filaments.

Tables 5.3 and Tables 5.4 show the AC loss reduction factor of the batches SAE-5 µm and SBE-10 µm, respectively. By comparing the values of these tables, one can observe that the AC loss reduction factor of both batches decrease with higher frequency. One can also observe that the AC loss reduction factor increases with higher number of filaments only at 6 Hz. At this frequency the coupling loss is not significant and therefore does not influence the AC losses.

The comparison of the AC loss reduction factor between the two batches for all frequencies indicates that the SAE-5 µm batch reduces the AC losses more efficiently than the other batch. This shows that the striation process plays a role in the reduction of the AC magnetization losses.

5.2.3 Influence of the Striation Process

For analyzing the influence of the striation process on the reduction of the AC magnetization losses, two striated tapes from the same batches with the largest difference of number of filaments were chosen. The 10-filament and 40-filament tapes were selected for comparison at two different frequencies, 6 Hz and 130 Hz. These two numbers of filaments were chosen

Table 5.5: AC loss reduction factor with respect to unstriated tape for the 10-filament and 40-filament samples at 6 Hz and 130 Hz, and at 50 mT.

Process	Copper Thickness*	Frequency	Number of Filaments	
			10 (8.7**)	40 (37.9**)
SAE	5 µm (8.6 µm)	6	7.8	20.6
		130	3.4	3.0
	10 µm (15.9 µm)	6	6.8	15.1
		130	2.5	2.5
SBE	5 µm (3.5 µm)	6	4.5	23.1
		130	1.0	4.0
	10 µm (8.1 µm)	6	3.2	4.5
		130	1.0	0.8

*First value is the expected copper thickness and the value in parenthesis is the measured thickness.

**In parenthesis, the Mawatari value corresponding to the number of filaments (10 and 40) and the applied field amplitude (50 mT).

because the 10-filament tapes are those with the smallest number of filaments and because three of the four 40-filament tapes at 24 Hz (Table 5.2) have the lowest AC losses.

Figure 5.6 shows the AC magnetization losses of all the 10-filament tapes for various frequencies (from 6 Hz to 130 Hz), alongside the theoretical Brandt and Mawatari curves. Table 5.5 reports the loss reduction values with respect to the unstriated tape for an applied field amplitude of 50 mT.

In Table 5.5, a further number is given (in parenthesis in the table): the expected loss reduction factor based on Mawatari's formula, for an applied field of 50 mT. This factor is similar, but not exactly equal, to the number of filaments, for two reasons: firstly, due to loss of superconducting material in the grooves, the filaments are narrower than the width of the unstriated tape divided by the number of filaments. Secondly, because the considered field (50 mT) is not sufficiently large to reach the high field limit where the loss reduction proportional to the number of filaments is to be expected.

All reference tapes follow the Brandt curve for field larger than 10 mT, but the data are not displayed for sake of readability.

- **10-filament tapes at** 6 Hz – The AC magnetization losses of the 10-filament tapes at 6 Hz are plotted in Fig. 5.6(a). The SAE tapes follow the Mawatari curve. The SBE tapes have higher loss level, due to coupling, which is more pronounced in the SBE-10 µm tape. At 6 Hz the SAE process reduces the AC magnetization loss more effectively than the SBE one for both copper thicknesses. In particular, for a field of 50 mT, the SAE-5 µm and SAE-10 µm have losses 7.8 and 6.8 times lower than the corresponding reference unstriated samples. The SBE-5 µm and SBE-10 µm have losses 4.5 and 3.2 times lower than the corresponding reference unstriated samples. These results are summarized in Table 5.5.

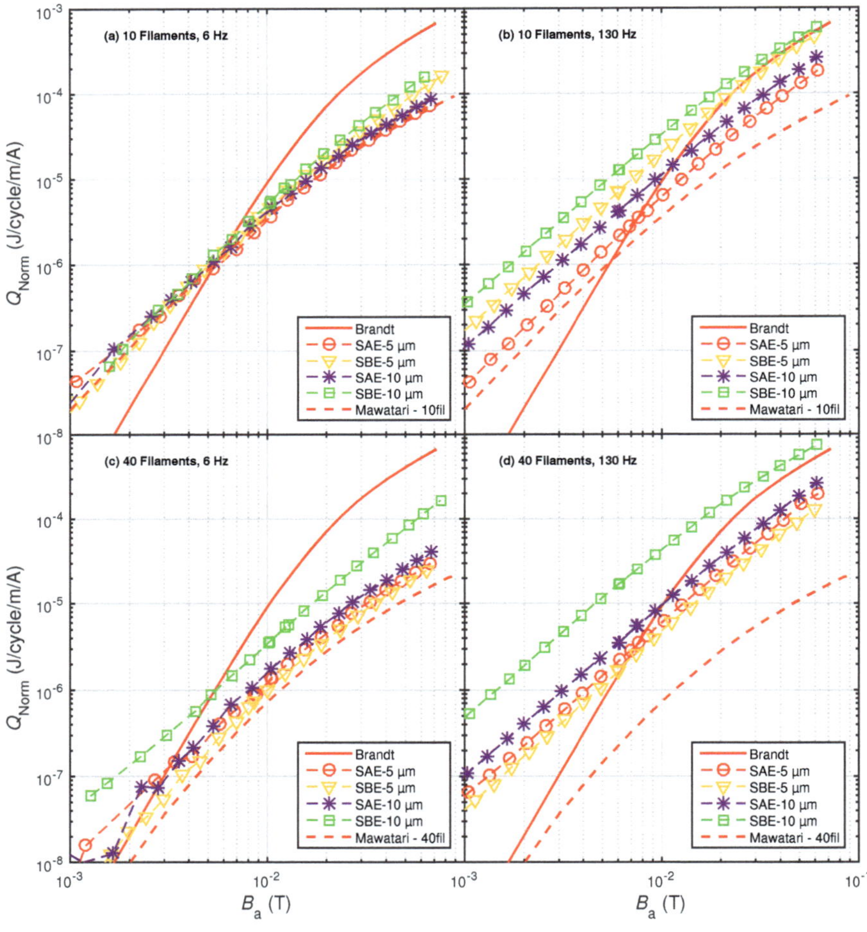

Figure 5.6: AC magnetization loss of the reference, 10-filament and 40-filament tapes of the all batches (SAE-5 µm, SAE-10 µm, SBE-5 µm and SBE-10 µm) at two frequencies. Samples as follows using different striation processes: (a) 10-filament tapes at 6 Hz, (b) 10-filament tapes at 130 Hz, (c) 40-filament tapes at 6 Hz and (d) 40-filament tapes at 130 Hz.

- **10-filament tapes at** 130 Hz – Figure 5.6(b) shows the AC magnetization losses of all the 10-filament tapes at 130 Hz. One can distinguish two patterns. The SBE tapes show almost no loss reduction with respect to the unstriated tape and, for fields above 20 mT, the data points are close to the Brandt curve. On the contrary, the SAE tapes show a more significant loss reduction and the data points are still far from the Mawatari limit. The losses of all the tapes are much larger than those at 6 Hz, because the coupling loss is proportional to the frequency.

- **40-filament tapes at** 6 Hz – Figure 5.6(c) shows the AC magnetization losses of the 40-filament tapes at 6 Hz. One can observe two patterns here. The AC losses of the SAE-5 µm, SAE-10 µm and SBE-5 µm tapes are close to the Mawatari limit. The AC losses of SBE-10 µm tape are much higher. As reported in Table 5.5, the corresponding loss reduction factors of the two patterns with respect to the unstriated tape are around 20 and 5, respectively. At this frequency and number of filaments, the thickness of the copper stabilization does not significantly influence the level of AC losses in SAE tapes and SBE tapes with smaller amount of copper. As observed previously, at such low frequency, the coupling loss should influence slightly the AC loss level, but it is not the case for the SBE-10 µm tape. This indicates a strong presence of a coupling loss in this case.

- **40-filament tapes at** 130 Hz – Figure 5.6(d) shows the AC magnetization losses of the 40-filament tapes at 130 Hz. The SBE-10 µm 40-filament sample has similar losses as the 10-filament tape in Fig. 5.6(c) with similar striation process and copper thickness: no loss reduction with respect to the unstriated tape. This is due to the fact that both samples are fully covered by the copper (as shown by the FIB cuts in Fig. 5.1(d)) and therefore have similar coupling loss level.

As far as the other types of 40-filament samples are concerned, they behave similarly to each other, which means that the tapes have the same coupling loss level. The values in Table 5.5 indicate that no process or amount of copper is preferable, except for the SBE-10 µm tape, which has larger losses (about three times higher than the other three).

5.3 Coupling Current Contribution to AC Losses

In order to support the behavior of the AC magnetization losses of the different tapes, the transverse resistivity of the coupling current were calculated from AC losses and compared to the transverse resistivity measured from direct transverse resistance measurements. This comparison was done with the 10-filament tapes.

5.3.1 Transverse Resistivity Calculation from AC Loss Measurements

In striated tapes there are three main contributions to AC losses. More details about those losses can be found in section 2.2.1. The hysteresis losses in the superconducting material are one of them. In this analysis, the filaments are considered to be electrically uncoupled and their losses are represented by Mawatari's formula as P_h.

Another loss contribution is given by eddy current losses, which are expected to occur in the normal metal parts of the tape, such as the Hastelloy, the copper layers and the silver layers. The eddy current losses per unit volume of a normal conductor slab are given by [Mül97]

$$P_{ec} = \frac{\pi^2}{6} \frac{B_a^2 f^2 W_{nc}^2}{\rho_{nc}} \tag{5.3}$$

where B_a, W_{nc} and ρ_{nc} represent the magnetic field amplitude, the width and the resistivity of the normal conductor, respectively.

The eddy current losses of the SAE-10 µm reference sample, which is the thicker sample, for $B_a = 50\,\text{mT}$ and $f = 130\,\text{Hz}$ are as follows: $\text{P}_{ec}(\text{Hastelloy}) = 8.72 \times 10^{-4}\,\text{W}$, $\text{P}_{ec}(\text{silver}) = 0.0138\,\text{W}$ and $\text{P}_{ec}(\text{copper}) = 0.1433\,\text{W}$. The dimensions used in the calculation are a width of 12 mm, a length of 0.15 m and a thickness of 60 µm for the Hastelloy, 2.3 µm for the silver and 15.9 µm for the copper. The resistivities used for the Hastelloy, the silver and the copper are 1.24 µΩ m, 3 nΩ m [AKY+04] and 1.9 nΩ m, respectively.

The comparison of the eddy current losses with the 10-filament Mawatari value, which is the lowest losses that the 10-filament sample can produce and that is equal to 0.316 W, shows that the eddy current in the Hastelloy and the silver layers can be neglected, but those in the copper layer have to be taken into account. In the case of SAE sample, which has two copper layers, the eddy current losses in the copper are estimated to be 0.2866 W, which is not very distant from the losses in the superconductor calculated with Mawatari's value. This means that if one wants to achieve the Mawatari value, the tape has to operate at low frequency in order to reduce those losses present in the copper.

The third contribution to the AC losses is the coupling losses. They occur due to the resistance between filaments. The formula of these losses per unit volume of the superconducting layer is given by [CO06]

$$P_c = \frac{(f B_a L)^2}{\rho_{tr}}. \tag{5.4}$$

The coupling loss can be deduced from the total AC losses

$$P = P_h + P_{ec} + P_c \tag{5.5}$$

and used to calculate the transverse resistivity ρ_{tr}. The latter is compared, later in this chapter, with the transverse resistivity calculated from the transverse resistance measurement of the 10-filament tapes.

The transverse resistivities evaluated from eq. (5.4) with the coupling loss calculated from eq. (5.5) are equal to 8.02 nΩm, 1.18 nΩm, 4.98 nΩm and 0.90 nΩm for the samples SAE-5 µm, SBE-5 µm, SAE-10 µm and SBE-10 µm, respectively.

5.3.2 Transverse Resistivity Measurement of the 10-filament Samples

The inter-filament resistance value can give information about the quality of the groove from an electrical point of view and therefore about the possible paths of the coupling currents.

In order to calculate the transverse resistance of the tape, a small DC current (200 mA) in the two outer filaments of 10-filament samples was injected and the voltage between a given filament and the outer filament used as reference was measured in a liquid nitrogen bath at atmospheric pressure (77 K). The resistance values were derived from the resulting current-voltage curves. The experimental setup and results are described in detail in [GGJ+16].

Figure 5.7 shows the inter-filament resistance values of the SAE and SBE tapes with different thicknesses of copper. Those values are compared to the resistances of the Hastelloy and

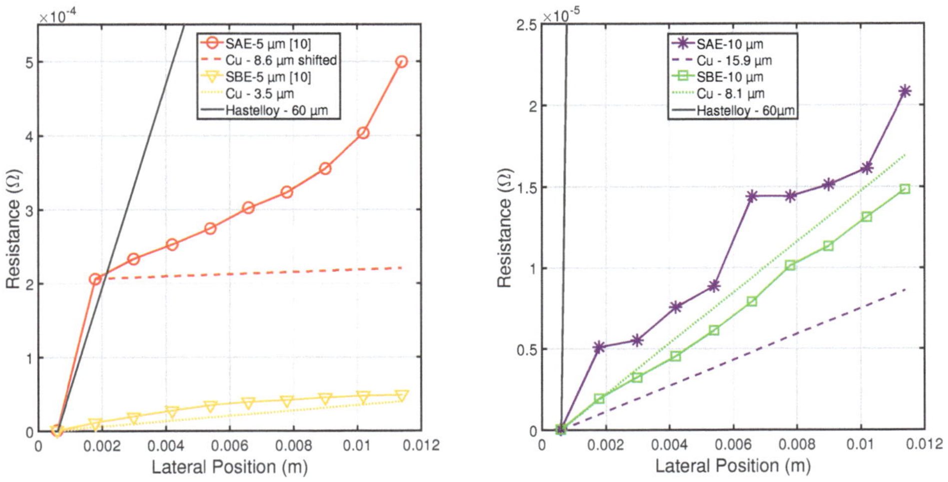

(a) (b)

Figure 5.7: Transverse resistance of 10-filament SuperOx tapes for the four striation processes. (a) SAE-5 µm and SBE-5 µm replotted from [GGJ+16], (b) SAE-10 µm and SBE-10 µm.

the copper, assuming electrical resistivities of 1.24 μΩ m [DVK+14] and 1.9 nΩ m [MCdS+14], respectively, thicknesses of the copper layer as in Table 5.1, and 150 mm long samples.

For the SAE-5 μm tape (Fig. 5.7(a)), the resistance value of the first point has a big jump, which corresponds to the barrier resistance between the superconducting layer and the Hastelloy. The slope of the points in the middle of the graph indicates where the current tends to flow. In this case the slope is between that of the copper and that of the Hastelloy. As explained in [GGJ+16], this indicates that the current splits into two paths: the first one where the current flows into the barrier resistance and then the superconducting layer, and the second path where the current flows into the Hastelloy.

Regarding the SBE-5 μm tape (Fig. 5.7(a)), the transverse resistance is close to that of the 3.5 μm copper layer. In this sample, the current flows through copper, as the groove is partly covered by copper (Fig. 5.1(b) of this thesis and Fig. 6 of [GGJ+16]).

In the SAE-10 μm tape (Fig. 5.7(b)), the transverse resistance is far from the curve corresponding to the actual copper thickness (15.9 μm), because there is some redeposited material in the groove after laser striation (Fig. 5.1(c)), but not a complete copper coverage caused by the electroplating as in the case of the SBE-10 μm tape.

In the SBE-10 μm tape (Fig. 5.7(b)), the transverse resistance approximately follows the curve for the 8.1 μm thick copper. This indicates that the current flows through the copper stabilization, which completely covers the groove (Fig. 5.1(d)).

The transverse resistance resulting from a current flowing between two filaments is given by

$$R_{\mathrm{tr}} = \rho_{\mathrm{tr}} \frac{w}{L t_{\mathrm{br}}} \tag{5.6}$$

where w is the sum of the widths of one filament and of one groove, L the length of the sample and t_{br} the thickness of the barrier. The latter is equal to the thickness of the superconducting layer for the samples with 5 μm of stabilizer and equal to the depth of the groove for the samples with 10 μm of stabilizer. The depth of the groove is approximately equal to the thickness of the copper, the silver and the superconducting layers. The transverse resistance measurement of a sample with N filaments can be written as $(N-1)R_{\mathrm{tr}}$.

Equation (5.6) is used to extract the transverse resistivity from the measurement of the resistance in Fig. 5.7. The extracted values are equal to 10.38 nΩ m, 0.80 nΩ m, 5.68 nΩ m and 2.30 nΩ m for the samples SAE-5 μm, SBE-5 μm, SAE-10 μm and SBE-10 μm, respectively. The transverse resistivities evaluated in section 5.3.1 from the coupling loss are equal to 8.02 nΩ m, 1.18 nΩ m, 4.98 nΩ m and 0.90 nΩ m for the samples SAE-5 μm, SBE-5 μm, SAE-10 μm and SBE-10 μm, respectively.

The transverse resistivities of the transverse measurement agree rather well with those calculated from the coupling loss, except for the sample SBE-10 μm. The transverse resistance measurement of the sample SBE-10 μm is not representative of the coupling current flowing between filaments. This sample is fully covered by the copper and thus the measurement shown in Fig. 5.7(b) gives simply the resistivity of the copper.

5.4 Summary

In this chapter, the AC magnetization loss results of tapes prepared with two different striation processes (SAE and SBE), with two different thicknesses of copper stabilization and various numbers of filaments, up to 60, were presented.

For all samples, making filaments helps to reduce the AC losses. The AC loss reduction is higher with higher number of filaments. The 40-filament tapes have the largest loss reduction factors (for the SAE-5 μm, SBE-5 μm and SAE-10 μm batches).

However, the expected level of loss reduction (given by the Mawatari formula) is only obtained at low frequencies (6 Hz). In all the other cases the coupling losses are substantial, sometimes dominant. The presence of a path where the coupling currents can easily flow has been confirmed by SEM images of the grooves and by transverse resistance measurements.

Based on the results presented here, the loss reduction strongly depends on the striation process and on the copper thickness. Within each process (SAE or SBE), the thinner copper layer allows reducing more effectively the AC losses and limiting coupling. For a given number of filaments, no process is preferable, except for the 10-filament case, where the SAE process has always the larger loss reduction. This is because the grooves of the 10-filament SBE tapes are filled by copper: partially with the 5 μm stabilization, wholly with the 10 μm stabilization. This causes large coupling losses. The SBE-10 μm tapes have always the larger losses, for any number of filaments, because the grooves are fully covered by copper.

One has to note that the use of SAE tapes in applications can be hindered by delamination problems, which have been reported by other authors [KKL$^+$16]. Further investigations on this potential problem are therefore necessary. But considering the complexity and duration of the two processes and the AC loss results shown here, the SAE process seems to be the preferable option.

The transverse resistivities evaluated from AC losses agree with those directly measured using DC current, except for the SBE-10 μm sample, where the presence of a copper layer over all the sample prevents a direct comparison.

6 AC Magnetization Losses of HTS High Current Cables

In this chapter the applicability of striations as means to reduce AC losses in HTS cables is investigated, with focus on CORC® cables. The CORC® cables were assembled by Advanced Conductor Technologies [Tec] with tapes from SuperPower [Supc].

This chapter is divided in three sections: Section 6.1 shows the investigation of the AC magnetization losses of the tapes that were used for assembling the CORC® cables. Based on the results presented in chapter 5, superconducting tapes with 5 µm-Cu stabilization were selected and the Striated After Electroplating (SAE) process was used for making the striations. Section 6.2 shows the AC losses of CORC® cables. Section 6.3 shows AC loss comparisons between CORC® and RACC cables. Finally, section 6.4 provides a summary of the results of this chapter.

In this chapter, two types of AC losses are shown: the AC magnetization losses, denoted as $Q_{\mathrm{AC,m}}$ (for tapes and cables), and the AC losses caused by the simultaneous presence of an AC transport current and an AC background field, denoted as $Q_{\mathrm{AC,tm}}$ (for cables).

6.1 AC Losses of 4 mm Wide HTS Tapes

Seven SuperPower tapes were prepared, six striated tapes (from 3 filaments to 30 filaments) and one non-striated tape, which was used as reference. Section 3.1 details the preparation of these tapes. A width of 4 mm was preferred to 12 mm for three reasons: firstly, by using three 4 mm wide tapes, one can expect to have 3 times lower magnetization losses than with one 12 mm wide tape (the 4 mm wide tapes can be viewed as a 12 mm wide tape striated into 3 filaments). Secondly, for a given former radius, the 12 mm wide tape undergoes a larger strain than the 4 mm wide tape. This strain can significantly lower the critical current of the tapes, when they are assembled in the cable structure [vdLE07, vdL09, vdLWN+16]. Thirdly, using narrower tapes also allows the possibility of a smaller former radius, thus increasing the cable's engineering current density.

The investigation of these single tapes consists in analyzing their properties such as the critical current and the AC magnetization losses, in order to find the appropriate tape candidate for assembling CORC® cables.

6.1.1 Tape Properties

The critical current of the striated tapes were measured before and after making the striations. The critical current of the tapes before striation is 103 ± 1 A. The value 103 A is taken as representative of the unstriated tapes.

Figure 6.1 shows the critical current of the striated tapes normalized by the corresponding critical current measured before striating them and plotted as a function of the number of

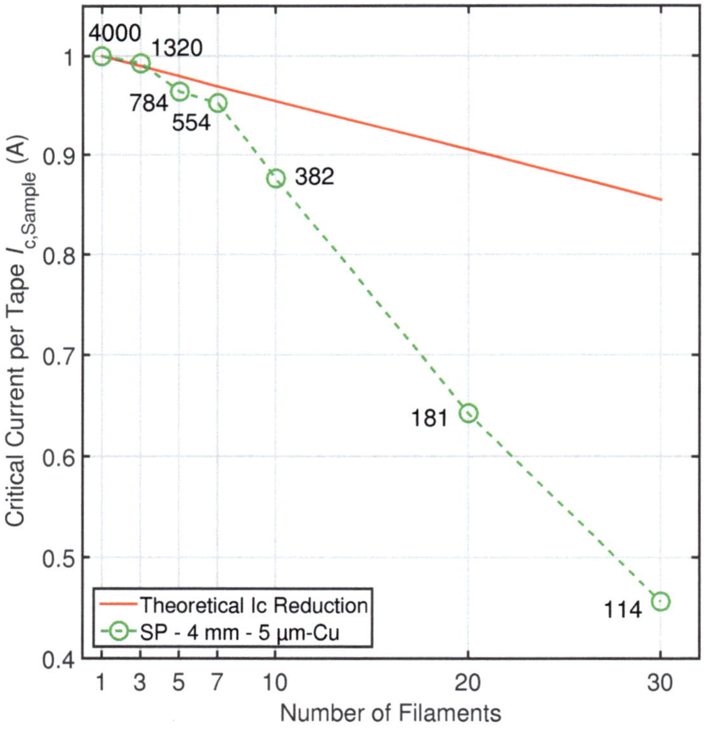

Figure 6.1: Critical current of 4 mm wide tapes normalized by the corresponding critical current measured before striating them. The continuous line represents the theoretical critical current that should be expected in striated tapes, based solely on the loss of superconducting material. The number next to each data point indicates the width of the filaments (µm).

filaments. The number next to each data point indicates the width of the filaments in µm. The continuous line represents the theoretical critical current that should be expected in striated tapes, based solely on the loss of superconducting material.

The critical current reduction caused by striation of the tapes is not significant only for tapes with 3 filaments, 5 filaments and 7 filaments, for which the reduction is 0.7 %, 3.5 % and 4.7 %, respectively. In tapes with 10 filaments, 20 filaments and 30 filaments, the reduction is 12.4 %, 35.7 % and 54.4 %, respectively. These reduction values were calculated with respect to the critical current measured before making the striations. The tapes with 10 filaments, 20 filaments and 30 filaments exhibit some degradation of their transport capability: in Fig. 6.1, this can be seen by the fact that the data points are far from the theoretical line. The tapes with 3 filaments, 5 filaments and 7 filaments are therefore the most suitable candidates for assembling the CORC® cables.

For assembling the CORC® cable, tapes with 5 filaments were chosen, so that the CORC® cables could be compared with previously built RACC cables, assembled from tapes with 5 filaments [KVK+16].

Figure 6.2 shows a Scanning Electron Microscope cross section picture of the Focused Ion Beam cut of one of the grooves of the 5-filament tape. One can observe that the filaments are well separated by a groove that is cut down to the Hastelloy layer. The width of this groove is around 20 µm. From this cross section picture, one can expect the coupling loss of such a striated tape to be small, because the coupling current can only flow through the Hastelloy, which has a large resistivity (1.24 µΩ m).

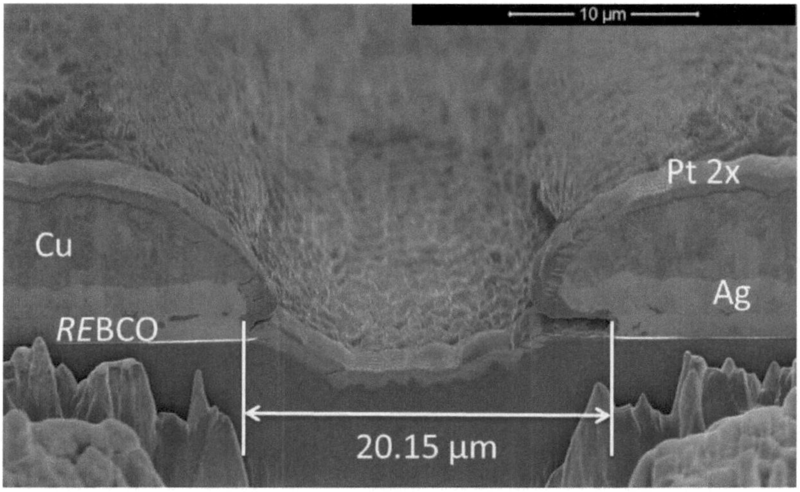

Figure 6.2: Scanning Electron Microscope cross section picture of the Focused Ion Beam cut of one of the grooves of the 5-filament tape.

6.1.2 AC Magnetization Losses of HTS Tapes

The measurement of the AC magnetization losses was carried out with the calibration-free method at 77 K. This method of measurement is detailed in section 3.1.2. In the following, the influence of the number of filaments and of the frequency on AC losses is investigated.

Influence of the Number of Filaments

Figure 6.3 shows the AC magnetization losses (normalized with respect to the critical current) as a function of the applied field amplitude at various frequencies. A loss reduction factor (formula 5.2) at 50 mT was calculated for each striated tape and reported in Table 6.1.

As already observed for the samples measured in chapter 5, the AC magnetization losses of the reference (non-striated) tapes at fields lower than 10 mT are larger than those predicted by Brandt's formula [BI93]. This happens for all frequencies, it has been reported in the literature and it is commonly ascribed to the non-uniformity of the sample's properties at the edges [AMM+06, GBM07, SPŠ+13].

The AC losses of the striated tapes at different frequencies are as follows:

- SP-4 mm tapes at 24 Hz – The AC losses decrease with higher number of filaments as shown in Fig. 6.3(a) and in Table 6.1. At this frequency, the background noise for a field larger than 20 mT is of the same order of magnitude than the AC losses of the 30-filament tape. That is why the AC loss curve is a bit wavy. At this low frequency the lower limit of the measuring system can be achieved by striated tapes, which have significantly reduced losses.

- SP-4 mm tapes at 72 Hz – The AC losses decrease with higher number of filaments up to 20 filaments as shown in Fig. 6.3(b) and in Table 6.1. The 20-filament and 30-filament tapes have similar AC losses.

- SP-4 mm tapes at 130 Hz – The AC losses decrease with higher number of filaments up to 20 filaments as shown in Fig. 6.3(c) and in Table 6.1. The AC losses of the 30-filament tape are between those of the 7-filament and the 10-filament tapes up to 40 mT, then between those of the 5-filament and the 7-filament tapes up to 60 mT, and between those of the 3-filament and the 5-filament tapes at higher fields.

As already observed for the samples measured in chapter 5, the reduction of the AC losses for the striated tapes starts at fields of a few mT (3-5), which are indicated by arrows in Fig. 6.3. At lower fields the losses of the striated tapes are higher than those predicted by Brandt's formula. This is because in the case of striated tapes the field penetrates into a larger volume of superconductor than in the non-striated case, where the center of the tape is more effectively screened from field penetration [Maw96, MZX+09, TVG+11].

From Fig. 6.3, one can note that, in general, the AC losses decrease with higher number of filaments up to 20 filaments and that the AC losses of the 30-filament tape increase with

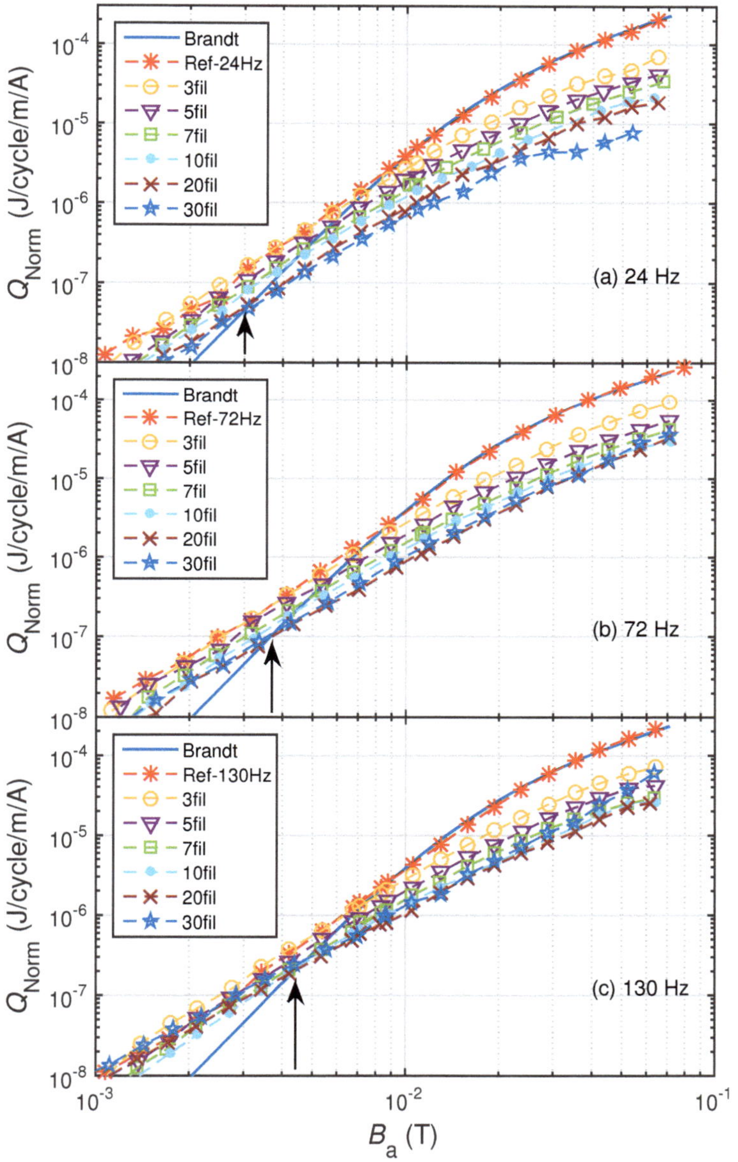

Figure 6.3: AC magnetization losses of the SuperPower 4 mm wide tapes with respect to the applied field amplitude at various frequencies: (a) 24 Hz, (b) 72 Hz and (c) 130 Hz. Black arrows are also drawn to indicate the start of the reduction of the AC losses for the striated tapes.

Table 6.1: AC loss reduction factor of all striated tapes with respect to unstriated tape at various frequencies (24 Hz, 72 Hz and 130 Hz) and at 50 mT.

Frequency	Number of Filaments*					
	3 (2.6)	5 (4.5)	7 (6.4)	10 (9.3)	20 (20.2)	30 (32.3)
24 Hz	3.3	4.6	6.3	8.1	9.1	20.3
72 Hz	2.4	4.1	5.4	6.6	7.4	6.6
130 Hz	2.6	4.0	6.3	6.4	6.8	4.3

*The first value is the number of filaments and the value in parenthesis is the reduction factor (calculated with Mawatari's formula) corresponding to the associated number of filaments and the applied field amplitude (50 mT).

higher frequency. This can be interpreted by the presence of a coupling loss. The increase of the coupling losses with higher number of filaments has also been observed in chapter 5 and in [DVK$^+$14].

Influence of the Frequency

In order to observe the influence of the frequency on the AC magnetization losses, one can compare the values of the reduction factor in Table 6.1 (vertical reading). The reduction factor values of the striated tapes between 3 filaments and 7 filaments are similar for the three frequencies. On the other hand, the reduction factor values of the striated tapes between 10 filaments and 30 filaments decrease with higher frequencies. This indicates that the latter tapes have in addition a coupling loss that the tapes up to 7 filaments do not have. An example of AC losses of the 5-filament and 20-filament tapes at all frequencies is shown in Fig. 6.4.

Figures 6.4(a) and 6.4(b) show also the Mawatari limit (dashed line), which represents the lowest limit that a striated tape can achieve when only hysteresis loss is considered. At fields larger than 20 mT, the losses of the 5-filament tapes follow the Mawatari limit, for all frequencies (Fig. 6.4(a)). This means that the coupling losses are negligible and therefore the filaments can be considered to be not coupled. On the contrary, the tapes shown in Fig. 6.4(b) are slightly separated from each other (the values in Table 6.1 show better this difference) and are well above the Mawatari curve. On the other hand, when the AC magnetization losses of Mawatari are normalized by the actual critical current of the tape, then this new limit (light blue curve in Fig. 6.4(b) is closer to the AC magnetization losses of the 20-filament tape at 24 Hz, but still lower than the latter. So, in this case, coupled filaments are the reason of this loss increase.

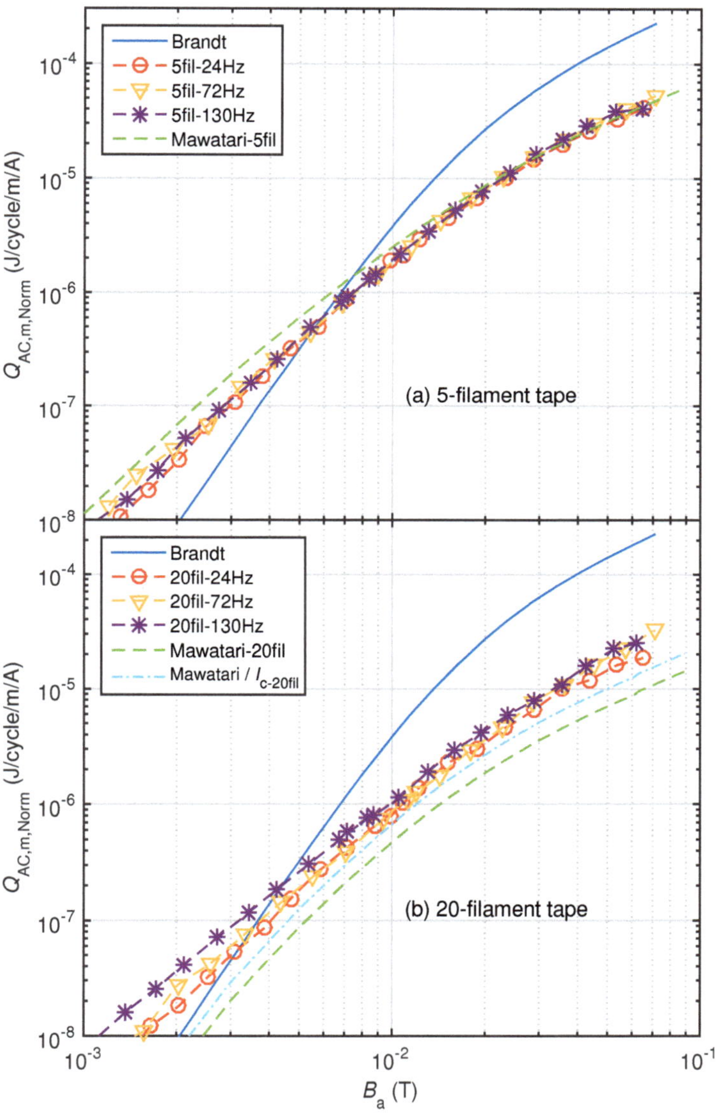

Figure 6.4: Normalized AC magnetization losses by the critical current of two striated SuperPower 4 mm wide tapes with respect to the applied field amplitude at various frequencies (24 Hz, 72 Hz and 130 Hz): (a) the 5-filament tape, the Mawatari limit for a tape with 5 filaments is also shown (green dashed line) and (b) the 20-filament tape, the Mawatari limit for a tape with 20 filaments as well as the Mawatari limit using the critical current of the actual tape are also shown (green and light blue dashed lines, respectively).

6.2 AC Losses of HTS CORC® Cables

As detailed in section 3.2.2, six CORC® cables were prepared with three different number of layers (2 layers, 3 layers and 4 layers). For each number of layers one cable was assembled with non-striated tapes used as reference and one cable was assembled with striated tapes. Tapes with 5 filaments were used for assembling the striated CORC® cables.

In this section the influence of striations as means to reduce the AC losses of CORC® cables and the influence of the number of layers on AC losses are investigated. The AC loss measurements were carried out at 77 K and at 130 Hz with the calorimetric method described in section 3.2.2.

In order to show the influence of striations, two types of AC losses are measured: the AC magnetization losses $Q_{AC,m}$ and the AC losses caused by the simultaneous presence of an AC transport current and an AC magnetic field $Q_{AC,tm}$.

The AC magnetization losses of the cables with 2 layers and 3 layers were too low to be measured by the calorimetric method. The non-striated reference cables assembled with 2 layers and 3 layers showed severe irreversible degradation while measuring them. That is why only the AC losses of the non-striated reference 4-layer CORC® cable and all striated CORC® cables with all layers were measured.

In order to show the influence of the number of layers, the AC losses $Q_{AC,tm}$ of the striated CORC® cables with different number of layers were measured. In this measurement the background field was chosen to be proportional to the transport current. This reproduces the condition usually found in coils, where the cable is subjected to the transport current and the magnetic field generated by the coil itself. The following ratios β of proportionality between field and current were used: 33 µT A^{-1}, 66 µT A^{-1} and 100 µT A^{-1}. If one takes as example the ratio 100 µT A^{-1} and an amplitude of current of 300 A carried by the cable, the corresponding external magnetic field applied to the cable is 30 mT.

6.2.1 Influence of Striations

AC Magnetization Losses ($Q_{AC,m}$)

Figure 6.5 shows the AC magnetization losses of the non-striated and striated 4-layer CORC® cables with respect to the amplitude of the applied magnetic field. The curves intersect at an applied field of 26 mT, below which the AC losses of the striated sample are higher than those of the non-striated sample. Above that field, the striated sample has lower AC magnetization losses. The behavior of the two AC loss curves is similar to that observed in section 5.2 and section 6.1.2 for single tapes, which is explained by the fact that in the case of striated tapes the field penetrates into a larger volume of superconductor than in the non-striated case, where the center of the tape is more effectively screened from field penetration [Maw96, MZX+09, TVG+11].

For an applied field of 50 mT and a frequency of 130 Hz, the reduction factor of a single striated tape with 5 filaments is 4 as shown in Table 6.1. Regarding the reduction factor of the CORC® cables, which is equal to 1.7, the striations reduce the AC magnetization losses but not as much as expected from a single tape. This lower reduction of the AC losses is probably due to the fact that the inner layers are actually shielded by the outer layers of the CORC® cable.

AC Losses of the Simultaneous AC Transport Current and AC Magnetic Field ($Q_{AC,tm}$)

Figure 6.6 shows the AC loss measurement of the simultaneous AC transport current and AC background field $Q_{AC,tm}$ of the non-striated and the striated 4-layer CORC® cables. The

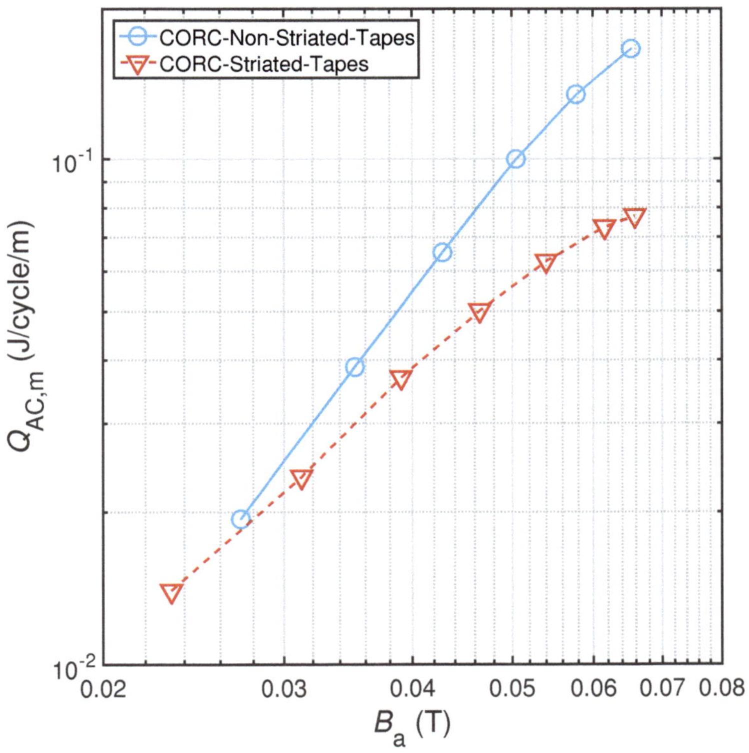

Figure 6.5: AC magnetization losses with respect to the amplitude of the applied magnetic field of the 4-layer CORC® cable with non-striated tapes and the 4-layer CORC® cable with striated tapes. These measurement were carried out at 130 Hz and at 77 K.

measurement of the AC transport losses, denoted as $Q_{\mathrm{AC,t}}$, of the non-striated cable is also plotted in order to get a rough estimation of their contribution to the total AC losses $Q_{\mathrm{AC,tm}}$.

When a background magnetic field is applied ($\beta \neq 0$) the AC losses of the striated samples intersect those of their corresponding non-striated samples for all field ratio β values, as observed in Fig. 6.3, Fig. 6.4 and Fig. 6.5, and section 5.2. The losses of the non-striated and of the striated cable increase with higher β values. This is expected because losses increase with higher applied fields B_{a}.

At an applied current of 800 A, the AC losses $Q_{\mathrm{AC,tm}}$ of the non-striated cables corresponding to a $\beta = 33\,\mathrm{\mu T\,A^{-1}}$ (blue cross marker line) are 3.5 times larger than the AC transport losses $Q_{\mathrm{AC,t}}$ (blue square marker line). This indicates that the contribution of the AC transport losses in the AC losses caused by the simultaneous AC transport current and AC background field is small. The contribution of the AC transport current is even smaller for the AC losses measured for a larger β.

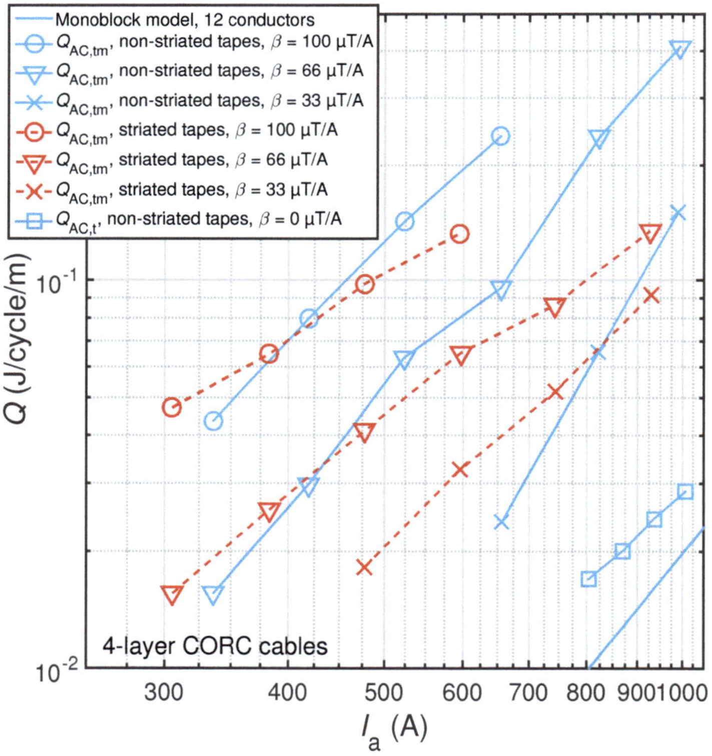

Figure 6.6: AC losses $Q_{\mathrm{AC,tm}}$ of the 4-layer non-striated and 4-layer striated CORC® cables for different background field amplitudes with respect to the amplitude of the applied current.

In this Fig. 6.6, the losses computed with the analytical monoblock model are also plotted in order to observe its adequacy with the AC transport losses. The expression of monoblock model in loss per unit length and per cycle is given as follows [VM95] and described in section 2.2.3:

$$Q_{\mathrm{monoblock}} = \frac{\mu_0}{2\pi}\frac{I_\mathrm{c}^2}{h^2}\{(2-Fh)Fh+2(1-Fh)\ln(1-Fh)\} \tag{6.1}$$

with $h = (D_\mathrm{c}^2 - D_\mathrm{f}^2)/D_\mathrm{c}^2$ and $F = I_\mathrm{a}/I_\mathrm{c}$. D_c is the outer diameter of the cable, D_f the outer diameter of the former, I_a the transport current amplitude of the cable and I_c the critical current of the cable.

Two reasons can be the cause of this difference: first, the fact that the other AC losses such as the eddy current and the transverse resistance between the tapes are not calculated and not considered in the model, which calculates only the hysteresis losses of the superconductors. Second, the fact that the model is not perfectly representative a CORC® cable. In particular, the model simulates a hollow cylinder, and the lateral gaps between the tapes are not taken into account. In the real cable, the magnetic field penetrates in these gaps, and is oriented perpendicularly to the flat face of the coated conductor. This can cause a substantial AC loss, which can be comparable or even higher than that calculated by the monoblock model.

Figure 6.6 shows that the AC transport losses $Q_{\mathrm{AC,t}}$ take a small share of the AC losses $Q_{\mathrm{AC,tm}}$. In order to know the real contribution of AC transport losses $Q_{\mathrm{AC,t}}$ and AC magnetization losses $Q_{\mathrm{AC,m}}$ to the AC losses $Q_{\mathrm{AC,tm}}$, the three types of losses are plotted in Fig. 6.7(a).

The losses of the 4-layer CORC® cable assembled with non-striated tapes are plotted as a function of the applied current in this figure. The AC losses $Q_{\mathrm{AC,tm}}$ measured with a background field of $\beta = 100\,\mu\mathrm{T\,A^{-1}}$ were selected. The applied field values corresponding the AC magnetization losses were converted to applied current with the ratio $\beta = 100\,\mu\mathrm{T\,A^{-1}}$ so that they fit to the figure. Regarding the AC transport losses, a fitting curve was calculated with a linear function on log-log scale (gradient equal to 2.34) in order to get the AC loss values corresponding to low applied current values. This fit is also plotted in Fig. 6.7(a). The sum of the AC magnetization $Q_{\mathrm{AC,m}}$ with the latter fit is plotted as well.

The sum is not equal to and is smaller than the AC losses $Q_{\mathrm{AC,tm}}$. The sum is very close to the AC magnetization losses $Q_{\mathrm{AC,m}}$ (5 % higher at 510 A, hence 51 mT). The gap between the losses is better observed with the zoom in Fig. 6.7(b). The AC magnetization losses are not that far from the AC losses caused by the simultaneous presence of an AC transport current and an AC background field, which is 30 % higher at 51 mT. Consequently, the share of the AC transport losses $Q_{\mathrm{AC,t}}$ in the AC losses $Q_{\mathrm{AC,tm}}$ is again observed to be small and the AC losses $Q_{\mathrm{AC,tm}}$ are mainly AC magnetization losses.

So, in order to get a rough estimation of the AC losses $Q_{\mathrm{AC,tm}}$, one can measure solely the AC magnetization losses $Q_{\mathrm{AC,m}}$. The AC magnetization losses are more advantageous to be measured because they much require less equipment, in terms of power sources, lock-in amplifiers, and Rogowski coils. In addition the AC magnetization losses can be measured by other techniques such as the calibration-free method shown in section 3.2.2. That brings more flexibility to such a measurement.

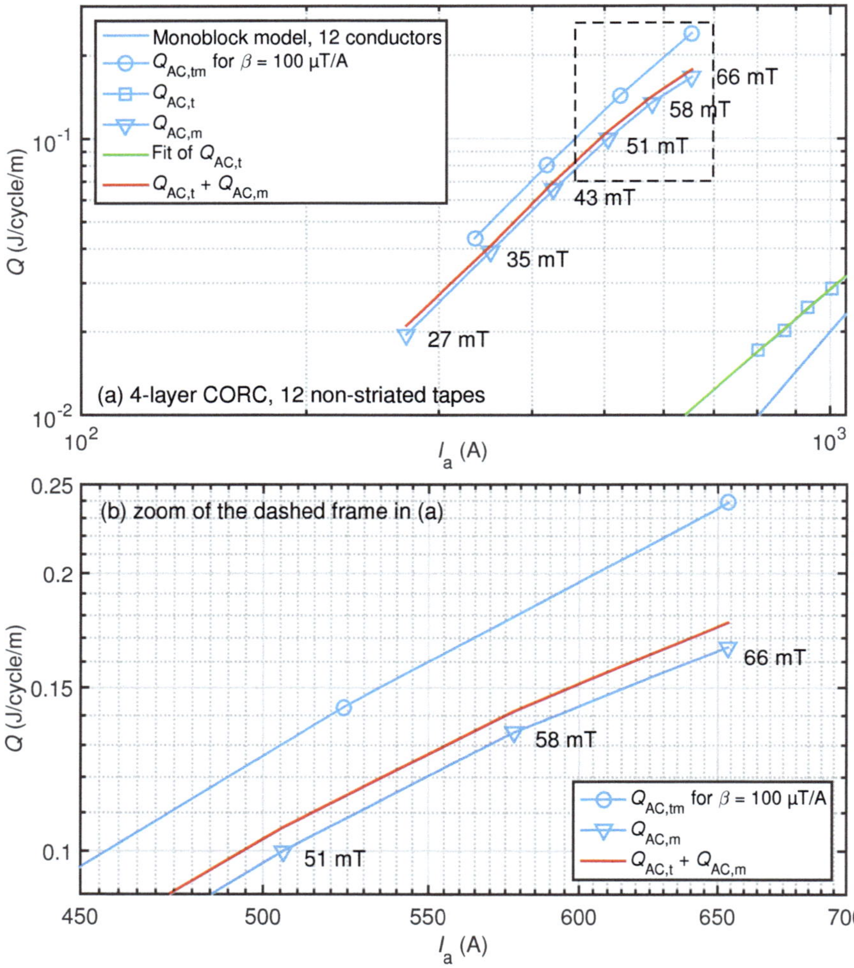

Figure 6.7: $Q_{\mathrm{AC,tm}}$ (ratio: 100 µT A^{-1}) of the 4-layer non-striated CORC® cable with respect to the amplitude of the applied current: (a) the AC magnetization losses and the AC transport losses of the this cable are also shown. A fitting curve was calculated with a linear function on a log-log scale for visualizing the AC transport losses at low current amplitudes in order to compare it with the AC magnetization losses $Q_{\mathrm{AC,m}}$ and (b) zoom in order to better see the difference between the AC losses $Q_{\mathrm{AC,tm}}$ and the sum of the AC magnetization losses $Q_{\mathrm{AC,m}}$ with the AC transport losses $Q_{\mathrm{AC,t}}$.

6.2.2 Influence of Layers

The AC losses $Q_{\mathrm{AC,tm}}$ of the three striated CORC® cables with respect to the various field ratios β are plotted in Fig. 6.8 as a function of the amplitude of the applied current. These cables differ in terms of the number of layers (2 layers, 3 layers and 4 layers shown in (a), (b) and (c), respectively of Fig. 6.8).

For a given CORC® cable, the gradient of the AC loss curves changes with the value of the ratio β. The gradient becomes lower with higher β values. This was also observed with the AC magnetization losses (Fig. 6.3 for single tapes and Fig. 6.5 for cables), which sees the value of their gradient changed to be smaller with higher applied magnetic field. So, the gradient is smaller when the applied field approaches the saturation.

Also for a given cable, one can observe that if the AC losses, corresponding to the three various β values, continue beyond the critical current of their corresponding cable, which are represented by the fitting full lines in Fig. 6.8, then the AC loss curves measured at β equal to 33 μT A^{-1} and 100 μT A^{-1} intersect each other. The crossover points are shown by vertical black arrows in the figure. When the AC loss curve measured at β equal to 66 μT A^{-1} is not too wavy, as shown in Fig. 6.8(c), then it is close to intersect the two others at the same crossover point. This latter is equal to 830 A, 1680 A and 3920 A for the 2-layer, 3-layer and 4-layer CORC® cables, respectively. The value of these crossover points are equal approximately to number of layers times the critical current of their corresponding CORC® cables, which are 410 A, 588 A and 963 A for respectively, 2 layers, 3 layers and 4 layers. As explained for a single tape in [DGZG16], the critical current can be deducted from the AC transport losses of a current beyond of the critical current of the sample, which is partly the case here because AC losses $Q_{\mathrm{AC,tm}}$ is caused by the simultaneous AC transport current and AC background field.

Consequently, the calculation of the critical current from AC loss measurement can be done instead of the regular DC characterization. Using AC current can be beneficial for two reasons: first, the use of AC current causes less heat dissipation than DC current at the same amplitude. An excessive heat dissipation can damage the sample (tape or cable) when this latter do not have a good thermal stability. Second, the flexibility of supplying AC current. For example, for supplying AC current into tape or cable, the same AC power source can be used, helped by a transformer in order to increase the current.

The difference between the critical current of the CORC® cables is large. The critical currents of the CORC® cables are 410 A, 588 A and 963 A for 2 layers, 3 layers and 4 layers, respectively. That is why the AC losses of the three striated CORC® cables were normalized by the critical current in Fig. 6.9.

One can observe that the AC losses of a given CORC® cable is equal to those of a cable with one more number of layer when the field decreases (smaller β) and one less number of layers when the field increases (higher β). This happens for example for the AC losses of the 2-layer CORC® cable (measured at a β equal to $100\,\mu\mathrm{T\,A^{-1}}$) that are similar to those of the 3-layer and 4-layer CORC® cables (measured at a β equal to $66\,\mu\mathrm{T\,A^{-1}}$ and $33\,\mu\mathrm{T\,A^{-1}}$, respectively).

So, in a range of current within the critical current of the three CORC® cables and a given limit of losses, lowering the number of layers allows increasing the background field.

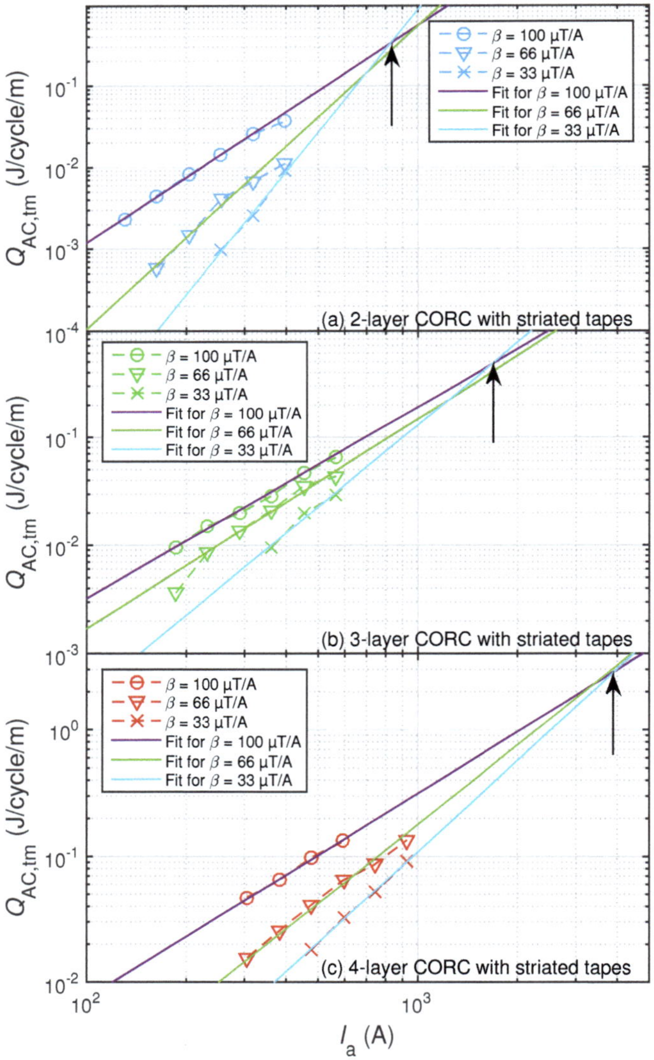

Figure 6.8: AC losses $Q_{AC,tm}$ of all striated CORC® cables for different background field amplitudes β as a function of the amplitude of the applied current: (a) AC losses of the 2-layer CORC® cable, (b) AC losses of the 3-layer CORC® cable and (c) AC losses of the 4-layer CORC® cable.

6.3 AC Loss Comparison Between HTS CORC® and HTS RACC Cables

In this section the AC losses of the CORC® cables investigated in section 6.2 are compared to the AC losses of their direct cable competitors, which are the RACC cables.

The section contains two investigations. Firstly, the AC losses of non-striated cables are compared in order to observe the applicability of the cable structure for reducing the AC losses. Secondly, striated cables are investigated in order to observe the applicability of the striations for better reducing the AC losses with respect to the non-striated cables.

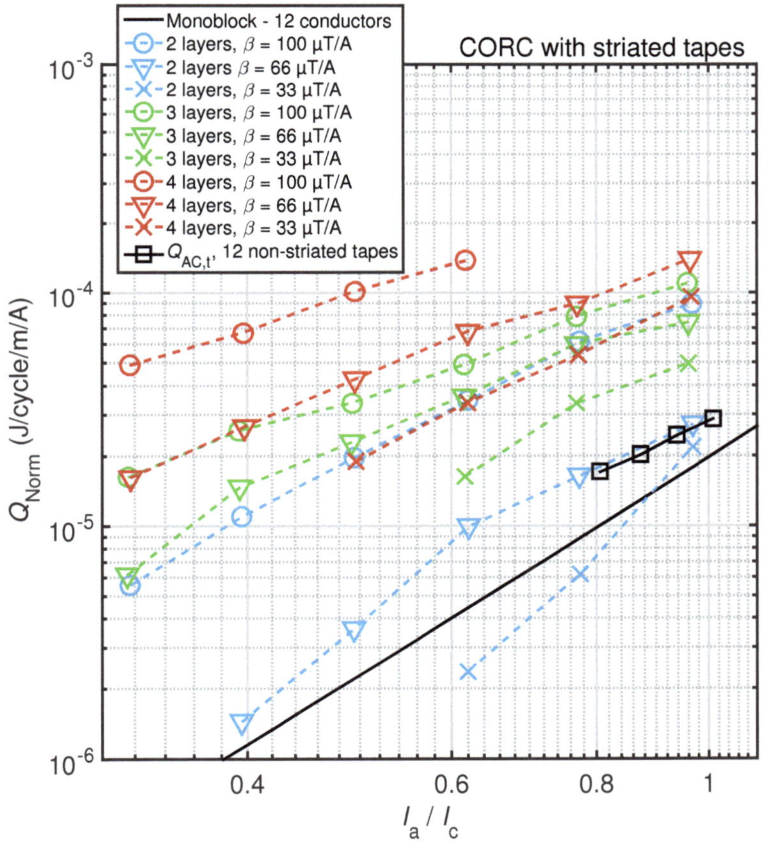

Figure 6.9: Normalized AC losses of the three striated CORC® cables with respect to the normalized transport current amplitude for different background field amplitudes β.

6.3.1 AC Losses of Non-Striated Cables

Cables having a similar critical current value, 998 A and 914 A were selected for this investigation for CORC® and RACC cables, respectively. The CORC® cable was assembled with

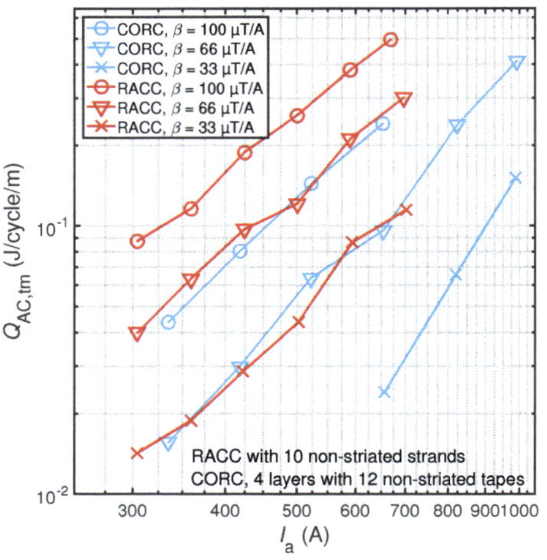

(a)

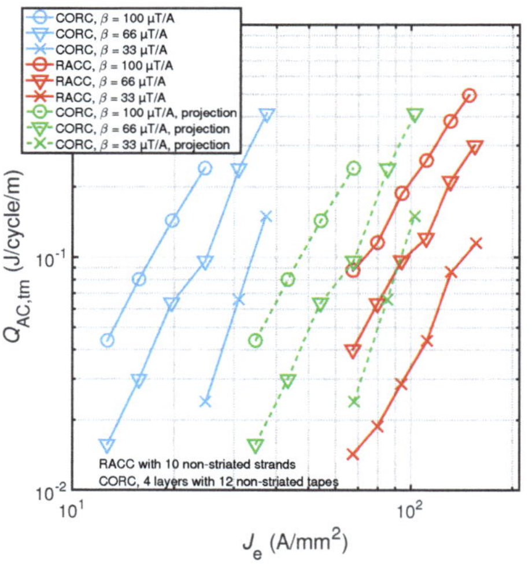

(b)

Figure 6.10: AC losses $Q_{\mathrm{AC,tm}}$ of the non-striated 4-layers CORC® cable (SP) and the RACC non-striated 10-strand cable (SOx) for different background field amplitudes: (a) with respect to the amplitude of the applied current I_{a} and (b) with respect to the amplitude of the engineering current density J_{e}. The AC loss data were taken from [KVK+16].

12 non-striated 4 mm wide tapes over 4 layers and the RACC was assembled with 10 non-striated 5.5 mm wide strands. Both were electroplated with 5 µm of copper for stabilization. The measurement of the AC losses $Q_{AC,tm}$ for various amplitudes of the background field is shown in Fig. 6.10.

The AC losses of the CORC® cable are 2 times lower for β equal to 100 µT A^{-1}, 3 times lower for β equal to 66 µT A^{-1}and 5 times lower for β equal to 100 µT A^{-1} than those of the RACC cable for a given applied current as observed in Fig. 6.10(a).

This difference comes from the shape of the cable and the position of the cable strands with respect to the applied field. Regarding the RACC cable, the field is applied perpendicularly to the cable strands, which is the the orientation for which the AC losses are the highest. In contrast, the tapes of the CORC® cable experience a field of various orientation, varying from perpendicular to parallel. Those different positions of strands with respect to the field produce AC losses between the perpendicular and parallel cases. In addition the CORC® cable has a smaller twist pitch, which can also help to reduce the losses.

Figure 6.10(b) shows the same AC losses $Q_{AC,tm}$ than in Fig. 6.10(a), but now plotted as a function of the engineering current density of the cables. For a given ratio of background field and an AC loss level, the RACC cable has a larger engineering current density. This can be explained by the fact that the tapes of the CORC® cables are wound around a former, which lowers the engineering current density. In CORC® cable, a raise of this current density can be obtained by reducing the diameter of the former.

As shown in [vdLWN+16], a former diameter of 3.2 mm was used for assembling a CORC® cable with 4 mm wide tapes and 50 µm thick substrate without significant degradation of the critical current. By assuming the same tapes used in this manuscript and no degradation of the critical current, the new engineering current density calculated with the smaller former is shown in Fig.6.10(b) (dashed lines). Smaller former can be used with tape having smaller width and substrate. So, in term of AC losses and current density, CORC® cable with smaller former can be a good candidate.

6.3.2 AC Losses of Striated Cables

Figure 6.11 shows the AC losses of the two RACC cables and the two CORC® cables measured with a background field of 100 µT A^{-1}. One with non-striated strands and one with striated strands (5 filaments) for RACC and CORC® cables. As also observed in section 3.2.1, the striated CORC® cable reduces the AC losses after the crossover point, which is 400 A. Regarding the RACC cables, the AC losses of the non-striated and striated cables have a similar level of AC losses and their gradient coefficient are also similar. Consequently the striations made on the RACC cable strands do not reduce the AC losses.

The fact that the striations do reduce the AC losses of a CORC® cable [ŠGK+13] and not significantly of a RACC cable is due to the filament geometry on the cable. The filaments in CORC® cables are transposed within each layer while those in RACC cables are not. If one takes one of the filaments that is located on one of the edges of the tape or strand for example, then this filament goes, at the half of the transposition length, to the opposite cable edge in the CORC® cable but not in a RACC cable.

6.4 Summary

The investigation of the AC losses of seven 4 mm wide and 5 µm thick copper tapes with various number of filaments showed that the fact of striating the tapes reduces the AC mag-

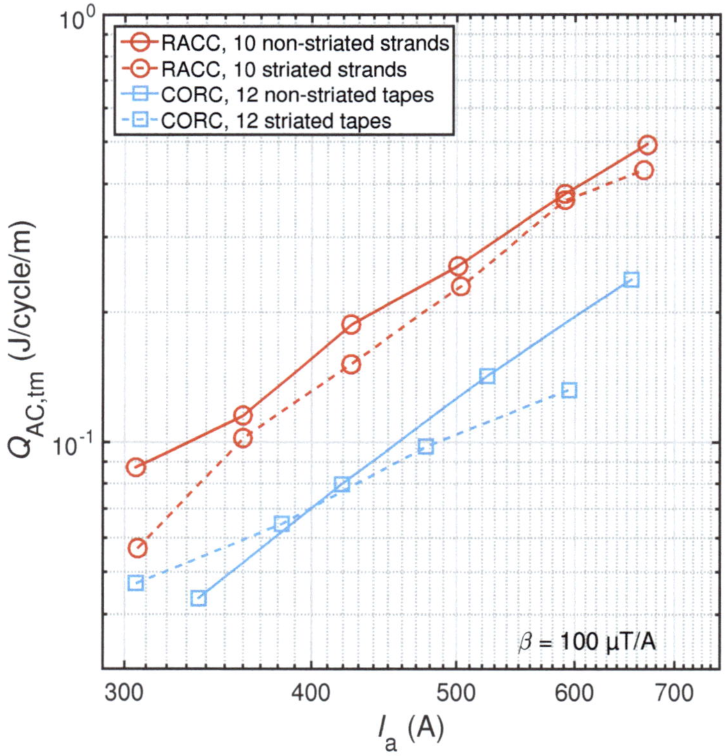

Figure 6.11: AC losses $Q_{AC,tm}$ of the non-striated and the striated 4-layers CORC® cables (SP), and the non-striated and the striated 10-strand RACC cables (SOx) for a background field amplitude of $\beta = 100\,\mu T\,A^{-1}$ with respect to the amplitude of the applied current I_a. The AC loss data were taken from [KVK+16].

netization losses. Those losses decrease with higher number of filaments until a certain limit, here 20 filaments. The most suitable candidates to be used for assembling a CORC® cables were the tapes with 3 filaments, 5 filaments and 7 filaments due to their small critical current degradation and their loss reduction proportional to the number of filaments.

The investigation of the AC magnetization losses $Q_{\mathrm{AC,m}}$ and the AC losses caused by the simultaneous presence of an AC transport current and an AC background field $Q_{\mathrm{AC,tm}}$ of two CORC® cables, one made with non-striated tapes and one made with 5-filament tapes, showed that the fact of striating the tapes helps to reduce both AC losses. The loss reduction of CORC® cables was smaller than that measured for single tape striated with 5 filaments (1.7 against 4 at 50 mT and 130 Hz). This smaller reduction can be explained by the fact that the inner layers are shielded by the outer layers.

The investigation of the AC losses $Q_{\mathrm{AC,tm}}$ of two CORC® cables (one non-striated and one striated) and two RACC cables (one non-striated and one striated), with similar critical current showed that for a given applied current the CORC® cables have smaller AC losses than those of the RACC cables. This can be explained by the fact that the strands of the RACC cable are exposed to a perpendicular magnetic field whereas those of the CORC® cable are exposed to a field of various orientations. When the current density is taken into account for the comparison of similar cables, the RACC cables have smaller AC losses. This can be explained by the fact that the former used in CORC® cable to wind the tapes reduces the engineering current density. In addition, if the former is reduced then the CORC® cable can become competitive.

7 Conclusion

The main goals of this thesis were: (i) investigate different strategies for obtaining uncoupled filaments in HTS coated conductors with copper stabilization: in particular, determine whether it is better to striate the tapes before of after depositing the copper layer; (ii) assess the applicability of the striation process to larger cables made of HTS coated conductor: in particular the CORC® and the RACC cables.

In order to get a consistent analysis of the effectiveness of making filaments on the AC magnetization loss reduction over the different striation processes and copper-stabilization thicknesses, four batches were realized with the same manufacturer (SuperOx) and the same width (12 mm). A batch consisted of one unstriated tape and four striated tapes from 10 filaments to 60 filaments, produced with different striation processes (SBE or SAE) and thicknesses of copper (5 µm or 20 µm).

The analysis of the AC losses was done on HTS striated tapes and striated cables, with a reference sample for both. For this, 27 tapes and 8 cables were prepared. The samples differ in technologies (with or without artificial pinning centers), in processing of striation (SBE or SAE), in thickness of the copper stabilizer (5 µm or 20 µm), in width (4 mm or 12 mm), in number of filaments (10 to 60), and in provider (SuperOx or SuperPower). The samples selected were striated using the laser ablation technique, which allows obtaining narrow filaments with precise cuts.

For all samples, making filaments helps to reduce the AC magnetization losses. The AC loss reduction is higher with higher number of filaments. The 40-filament tapes have the largest loss reduction factors, for the SAE-5 µm, SBE-5 µm and SAE-10 µm batches. However, the expected level of loss reduction, given by the Mawatari formula, is only obtained at low frequencies (6 Hz). At higher frequencies, the coupling losses are substantial, sometimes dominant. The presence of a path where the coupling currents can easily flow was confirmed by SEM images of the grooves and by transverse resistance measurements. Based on the results, the loss reduction strongly depends on the striation process and on the copper thickness. Within each process (SAE or SBE), the thinner copper layer allows reducing more effectively the AC losses and limiting coupling. For a given number of filaments, no process is preferable, except for the 10-filament case, where the SAE process has always the larger loss reduction. This is because the grooves of the 10-filament SBE tapes are filled by copper: partially with the

5 μm stabilization, entirely with the 10 μm stabilization. This causes large coupling losses. The SBE-10 μm tapes have always the larger losses, for any number of filaments, because the grooves are fully covered by copper.

One has to note that the use of SAE tapes in applications can be hindered by delamination problems, which have been reported by other authors [KKL+16]. Further investigations on this potential problem are therefore necessary. But considering the complexity and duration of the two processes and the AC loss results shown here, the SAE process seems to be the preferable option.

The transverse resistivities evaluated from AC losses agree with those directly measured using DC current, except for the SBE-10 μm sample, where the presence of a copper layer over all the sample prevents a direct comparison.

Based on the results obtained on the striation of tapes (presented in chapter 5), superconducting tapes with 5 μm-Cu stabilization and the SAE process for making the striations were selected to be analyzed in order to assemble CORC® cables. The investigation of the AC losses of seven 4 mm wide, SAE striation process, and 5 μm thick copper tapes with various number of filaments indicated that the most suitable candidates to be used for assembling a CORC® cables were the tapes with 3 filaments, 5 filaments and 7 filaments due to their small critical current degradation and their loss reduction proportional to the number of filaments.

The investigation of the AC losses caused by the simultaneous presence of an AC transport current and an AC background field of two CORC® cables, one made with non-striated tapes and one made with 5-filament tapes, confirmed that striating the tapes helps to reduce AC losses. The loss reduction of CORC® cables was smaller than that measured for single tape striated with 5 filaments (1.7 against 4 at 50 mT and 130 Hz). This smaller reduction can be explained by the fact that the inner layers are shielded by the outer ones.

The CORC® cable was then compared to its direct competitor, which is the RACC cable. These two cables have a different architecture, one is assembled by winding several layers of 4 mm wide tapes around a cylinder former, the other one consists of several meander-shaped 5.5 mm wide strands intertwined to each other. The number of tapes or strands defines the current-carrying capacity of the cable. Due to the different architectures, cables with similar critical current were chosen for the comparison. The investigation of the AC losses of two CORC® cables (one non-striated and one striated) and two RACC cables (one non-striated and one striated), with similar critical current showed that for a given applied current the CORC® cables have smaller AC losses than RACC cables. This can be explained by the fact that the strands of the RACC cable are exposed to a perpendicular magnetic field, whereas those of the CORC® cable are exposed to a field of various orientation. Once the losses are plotted as a function of the engineering current density, however, the RACC cables have smaller AC losses.

Designations and Abbreviations

$Ba-La-Cu-O$	Barium-Lanthanum-Copper-Oxide
CeO_2	Cerium Dioxide
CuO_2	Copper Oxide
$CuSO_4$	Copper Sulphate
H_2SO_4	Sulphuric Acid
M_gB_2	Magnesium Diboride
NbTi	Niobium Titanium
Nb_3Al	Niobium Aluminide
Nb_3Ge	Niobium Germanide
Nb_3Sn	Niobium Tin
$Ni-W$	Nickel-Tungsten
PbMoS	Lead Molybdenum Sulfur
YBCO	Yttrium Barium Copper Oxide
YSZ	Yttria-Stabilized Zirconia
Y_2O_3	Yttrium Oxide
$BaSnO_3$	Barium Stannate
$BaZrO_3$	Barium Zirconate

Gd_3TaO_7	Gadolinium Tantalum Oxygen
AC	Alternating Current
Ag	Silver
Al	Aluminum
AMSC	American Superconductor
AP	Advanced Pinning
BCS	Bardeen-Cooper-Schrieffer
CORC	Conductor On Round Core
Cu	Copper
CuNi44	Copper Nickel Alloys
DC	Direct Current
FIB	Focused Ion Beam
Ga	Gallium
Hg	Mercury
HTS	High-Temperature Superconductor
IBAD	Ion Beam Assisted Deposition
In	Indium
IR	Infrared
IRL	Industrial Research Limited
KIT	Karlsruhe Institute of Technology
LTS	Low-Temperature Superconductor
MOCVD	Metal Organic Chemical Vapour Deposition

MOD	Metalorganic Deposition
MRI	Magnetic Resonance Imaging
Nb	Niobium
Pb	Lead
RABiTS	Rolling Assisted Bi-axially Textured Substrate
RACC	Roebel Assembled Coated Conductor
REBCO	Rare Earth Barium Copper Oxide
SAE	Striated After Electroplating
SBE	Striated Before Electroplating
SC	Superconductor
SEM	Scanning Electron Microscopy
SMES	Superconducting Magnetic Energy Storage
Sn	Tin
Ta	Tantalum
Ti	Titanium
W	Tungsten
Zn	Zinc

List of Symbols

α	Reduction factor defined as the ratio between the normalized losses of the reference sample and the normalized losses of the striated sample	
β	Ratio of proportionality between the magnetic field and the transport current	(μT A^{-1})
κ	Ginzburg-Landau parameter	
λ	Penetration depth	(m)
μ_0	Permeability	(H m^{-1})
Φ_0	Quantum of flux	(T m^2)
ρ_{nc}	Resistivity of the normal conductor	(Ω m^{-1})
ρ_{tr}	Transverse resistivity of the striated tape	(Ω m)
ξ	Coherence length	(m)
a	Half width of the superconducting strip	(m)
B	Magnetic field	(T)
b	Damping parameter in $J_c(B,\theta)$ dependence	
$B_{\parallel}$	Magnetic flux density component, parallel to tape	
$B_{\perp}$	Magnetic flux density component, perpendicular to tape	
B_a	Magnetic field amplitude	(T)
B_{c1}	First critical magnetic field	(T)
B_{c2}	Second critical magnetic field	(T)
B_c	Critical magnetic field	(T)

B_c	Field-decay parameter in $J_c(B,\theta)$ dependence	
$B_c(0)$	Critical magnetic field at 0 K	(T)
d	Thickness of the superconducting strip	(m)
D_c	Outer diameter of the CORC® cable	(m)
D_f	Outer diameter of the former	(m)
e	Electronic charge	(C)
f	Frequency	(Hz)
h	Planck constant	(J s)
H_a	Magnetic field amplitude	(T)
I_a	Transport current amplitude	(A)
$I_{c,Sample}$	Critical current of the sample	(A)
I_{c1}	Critical current of the first criterion	(A)
I_{c2}	Critical current of the second criterion	(A)
I_c	Critical current	(A)
J	Current density	($A\,m^{-2}$)
J_c	Critical current density	($A\,m^{-2}$)
k	Anisotropy parameter in $J_c(B,\theta)$ dependence	
L	Length of the sample	(m)
N	Number of filaments of the superconducting tape	
n	Steepness of the $I-V$ curve	
P_c	Coupling losses in the striated tape due to the resistance between filaments	($J\,m^{-3}$)
P_{ec}	Eddy current losses of a normal conductor slab	($J\,m^{-3}$)
P_h	Hysteresis losses of a striated tape with electrically uncoupled filaments	($J\,m^{-3}$)
$Q_{AC,m}$	AC magnetization losses per cycle of the tape and the cable	($J\,m^{-1}$)

$Q_{AC,tm}$	AC losses per cycle of the cable caused by the simultaneous presence of an AC transport current and an AC background field	$(J\,m^{-1})$
$Q_{AC,t}$	AC transport losses per cycle of the cable	$(J\,m^{-1})$
$Q_{Norm,Ref}$	AC magnetization losses per cycle of the reference sample normalized by its critical current	$(J\,m^{-1}\,A^{-1})$
$Q_{Norm,Str}$	AC magnetization losses per cycle of the striated sample normalized by its critical current	$(J\,m^{-1}\,A^{-1})$
Q_{Norm}	AC magnetization losses per cycle normalized by the critical current of the sample	$(J\,m^{-1}\,A^{-1})$
Q_{Sample}	AC magnetization losses per cycle of the sample	$(J\,m^{-1})$
r	Wire radius	(mm)
R_c	Contact resistance of the cable copper leads	(nΩ)
R_{tr}	Transverse resistance between two filaments of the striated tape	(Ω)
T	Temperature	(K)
t_{Ag}	Silver thickness within the superconducting tape	(μm)
t_{br}	Thickness of the barrier resistance between two filaments of the striated tape	(m)
t_{Cu}	Copper thickness within the superconducting tape	(μm)
T_c	Critical temperature	(K)
t_{SC}	Superconductor thickness within the superconducting tape	(μm)
V_{c1}	Critical voltage of the first criterion	(V)
V_{c2}	Critical voltage of the second criterion	(V)
V_c	Critical voltage	(V)
w	Width of the filament	(m)
w_{groove}	Width of the space between two filaments	(m)
W_{nc}	Width of the normal conductor	(m)

Publications

Reviewed Journal Papers

[1] K. Kubiczek, F. Grilli, A. Kario, **A. Godfrin**, V. M. R. Zermeño, M. Stępień and M. Kampik, "Length Uniformity of the Angular Dependences of I_c and n of Commercial REBCO Tapes with Artificial Pinning at 77 K", *IEEE Transactions on Applied Superconductivity*, 29(1):8000309, Jan. 2019.

[2] **A. Godfrin**, A. Kario, R. Gyuráki, E. Demenčík, R. Nast, J. Scheiter, A. Mankevich, A. Molodyk, W. Goldacker, and F. Grilli, "Influence of the Striation Process and the Thickness of the Cu-Stabilization on the AC Magnetization Loss of Striated REBCO Tape", *IEEE Transactions on Applied Superconductivity*, 27(6):5900809, Sep. 2017.

[3] R. Gyuráki, **A. Godfrin**, A. Jung, A. Kario, R. Nast, E. Demenčík, W. Goldacker, and F. Grilli, "Interfilament Resistance at 77 K in Striated HTS Coated Conductors", *IEEE Transactions on Applied Superconductivity*, 26(8):6603606, Dec. 2016.

[4] E. Demenčík, **A. Godfrin**, V. M. R. Zermeño and F. Grilli, "Determination of I-V Curves of HTS Tapes from the Frequency-Dependent AC Transport Loss", *IEEE Transactions on Applied Superconductivity*, 26(3):9001004, Apr. 2016.

Presentations

[5] **A. Godfrin**, A. Kario, R. Gyuráki, E. Demenčík, R. Nast, J. Scheiter, A. Mankevich, A. Molodyk, F. Grilli and W. Goldacker, "AC loss analysis of striated *RE*BCO superOx coated conductors with Cu-stabilization", *Applied Superconductivity Conference (ASC 2016)*, Sep. 2016, Denver, Colorado (USA).

[6] A. Kario, M. Vojenciak, A. Kling, R. Nast, **A. Godfrin**, F. Grilli, D. van der Laan and W. Goldacker, "Comparison of AC losses in Roebel assembled coated conductor (RACC)

and CORC® cables assembled using filamentarised conductors at 77 K", *Applied Superconductivity Conference (ASC 2016)*, Sep. 2016, Denver, Colorado (USA).

[7] W. Goldacker, A. Kario, A. Kling, **A. Godfrin** and F. Grilli, "Comparison of different HTS AC cables for application in magnets, transformers and rotating machinery", *5th International Conference on Superconductivity and Magnetism (ICSM 2016)*, Apr. 2016, Fethiye (Turkey).

[8] W. Goldacker, A. Kling, A. Kario, F. Grilli, M. Vojenciak, **A. Godfrin**, B. Ringsdorf and R. Nast, "High current low AC loss HTS cable systems, an overlook and comparison of the concepts", *26th International Cryogenic Engineering Conference, International Cryogenic Materials Conference (ICEC-ICMC 2016)*, Mar. 2016, New Delhi (India).

[9] A. Kario, A. Kling, R. Nast, M. Vojenciak, **A. Godfrin**, B. Runtsch, F. Grilli, B. Ringsdorf, A. Jung and W. Goldacker, " Strand striation for reducing AC losses in Roebel Cables: is it a viable solution?", *26th International Cryogenic Engineering Conference, International Cryogenic Materials Conference (ICEC-ICMC 2016)*, Mar. 2016, New Delhi (India).

[10] F. Grilli, R. Gyuráki, **A. Godfrin**, A. Jung, A. Kario, R. Nast, E. Demenčík and W. Goldacker, "Transverse resistance and current paths in low-loss striated HTS coated conductors", *28th International Symposium on Superconductivity (ISS 2015)*, Nov. 2015, Tokyo (Japan).

[11] W. Goldacker, A. Kario, A. Kling, M. Vojenciak, **A. Godfrin**, B. Ringsdorf, R. Nast and D. van der Laan, "HTS-Roebel-cables in competition to the CORC approach", *12th European Conference on Applied Superconductivity (EUCAS 2015)*, Sep. 2015, Lyon (France).

[12] A. Kario, R. Nast, A. Jung, B. Ringsdorf, E. Demenčík, B. Runtsch, **A. Godfrin**, R. Gyuráki, A. Molodyk, A. Mankevich, J. Scheiter, F. Grilli and W. Goldacker, "Effectiveness of laser striation for AC loss reduction in SuperOx coated conductor", *Cryogenic Engineering Conference & International Cryogenic Materials Conference*, Jun. 2015, Tucson, Arizona (USA).

[13] F. Grilli, E. Demenčík, A. Kario, R. Nast, A. Jung, M. Vojenciak, **A. Godfrin**, V. M. R. Zermeno and W. Goldacker, "Striation of HTS Coated Conductors for AC loss Reduction: Underlying Mechanisms and Technological Feasibility", *27th International Symposium on Superconductivity (ISS 2014)*, Nov. 2014, Tokyo (Japan).

Posters

[14] R. Gyuráki, **A. Godfrin**, A. Jung, A. Kario, R. Nast, E. Demenčík, F. Grilli and W. Goldacker, "Measurement of inter-filament resistance in striated HTS coated conductors", *26th International Cryogenic Engineering Conference, International Cryogenic Materials Conference (ICEC-ICMC 2016)*, Mar. 2016, New Delhi (India).

[15] **A. Godfrin**, R. Gyuráki, E. Demenčík, A. Jung, A. Kario, R. Nast, J. Scheiter, A. Mankevich, A. Molodyk, F. Grilli and W. Goldacker, "AC magnetization loss and transverse resistance measurement of striated coated conductors with electroplated Cu-stabilization", *Superconducting Materials on their Way from Physics to Applications: 608. WE-Heraeus-Seminar*, Feb. 2016., Bad Honnef (Germany).

[16] **A. Godfrin**, R. Gyuráki, E. Demenčík, A. Jung, A. Kario, R. Nast, J. Scheiter, A. Molodyk, A. Mankevich, F. Grilli and W. Goldacker, "AC magnetization loss and transverse resistance measurement of striated coated conductors with electroplated Cu-stabilization", *12th European Conference on Applied Superconductivity (EUCAS 2015)*, Sep. 2015, Lyon (France).

[17] R. Nast, A. Kario, A. Jung, **A. Godfrin**, R. Gyuráki, B. Ringsdorf, J. Scheiter, F. Grilli, W. Goldacker, A. Molodyk and A. Mankevich, "Influence of electroporated Cu-stabilization on laser-structured Ag-cap coated conductors", *12th European Conference on Applied Superconductivity (EUCAS 2015)*, Sep. 2015, Lyon (France).

[18] A. Kario, M.Vojenciak, A. Kling, R. Nast, **A. Godfrin**, E. Demenčík, B. Ringsdorf, F. Grilli and W. Goldacker, "Magnetization loss of Roebel coated conductor cable with up to 20 filaments for AC applications", *12th European Conference on Applied Superconductivity (EUCAS 2015)*, Sep. 2015, Lyon (France).

[19] **A. Godfrin**, B. Ringsdorf, R. Nast, F. Grilli, W.1 Goldacker, "Mesure automatique de Ic(B,α) de rubans supraconducteurs HTc", *11èmes Journées de Cryogénie et de Supraconductivité*, May 2015, Aussois (France).

[20] E. Demenčík, **A. Godfrin**, F. Grilli, A. Kario, M. Vojenciak, R. Nast, A. Jung and W. Goldacker, "AC magnetization loss and transverse resistivity of striated YBCO coated conductors", *Applied Superconductivity Conference (ASC 2014)*, Aug. 2014, Charlotte, North Carolina (USA).

[21] **A. Godfrin**, B. Douine, K. Berger, S. Mezani, S. Bendali and J. Lévêque, "Détection de quench dans une bobine supraconductrice HTC à noyau de fer", *10èmes Journées de Cryogénie et de Supraconductivité*, Jun. 2012, Aussois (France).

Bibliography

[Abr57] A. A. Abrikosov. On the Magnetic Properties of Superconductors of the Second Group. *J, Exptl. Theoret. Phys. (U.S.S.R.)*, 32(32):1442–1452, Jun. 1957.

[Ain19] M. D. Ainslie. Reducing ac losses in high-temperature superconducting coated-conductor wires towards more efficient superconducting electric power applications. *Superconductor Science and Technology*, 32(3):030501, Jan. 2019.

[AKY$^+$04] N. Amemiya, S. Kasai, K. Yoda, Z. Jiang, G. A. Levin, P. N. Barnes, and C. E. Oberly. AC loss reduction of YBCO coated conductors by multifilamentary structure. *Superconductor Science and Technology*, 17(12):1174–1471, Dec. 2004.

[AMM$^+$06] N. Amemiya, O. Maruyama, M. Mori, N. Kashima, T. Watanabe, S. Nagaya, and Y. Shiohara. Lateral Jc distribution of YBCO coated conductors fabricated by IBAD/MOCVD process. *Physica C: Superconductivity and its Applications*, 445-448(1-2):712–716, Oct. 2006.

[ARZH$^+$11] M. D. Ainslie, V. M. Rodriguez-Zermeno, Z. Hong, W. Yuan, T. J. Flack, and T. A. Coombs. An improved FEM model for computing transport ac loss in coils made of rabits YBCO coated conductors for electric machines. *Superconductor Science and Technology*, 24(4):045005, Jan. 2011.

[AYK$^+$05] N. Amemiya, K. Yoda, S. Kasai, Z. Jiang, G. A. Levin, P. N. Barnes, and C. E. Oberly. AC Loss Characteristics of Multifilamentary YBCO Coated Conductors. *IEEE Transactions on Applied Superconductivity*, 15(2):1637–1642, Jun. 2005.

[Bea62] C. P. Bean. Magnetization of hard superconductors. *Physical Review Letters*, 8(6):250–252, 1962.

[BI93] E. H. Brandt and M. Indenbom. Type-II-superconductor strip with current in a perpendicular magnetic field. *Physical Review B*, 48(17):12893–12906, Nov. 1993.

[CBH+02] C. B. Cobb, P. N. Barnes, T. J. Haugan, J. Tolliver, E. Lee, M. D. Sumption, E. W. Collings, and C. E. Oberly. Hysteretic loss reduction in striated YBCO. *Physica C: Superconductivity*, 382(1):52–56, Oct. 2002.

[CO99] W. J. Carr and C. E. Oberly. Filamentary YBCO conductors for AC applications. *IEEE Transactions on Applied Superconductivity*, 9(2):1475–1478, Jun. 1999.

[CO06] W. J. Carr and C. E. Oberly. Transverse resistivity of a filamentary coated conductor. *Superconductor Science and Technology*, 19(1):64–67, Jan. 2006.

[CSZ+09] Y. Chen, V. Selvamanickam, Y. Zhang, Y. Zuev, C. Cantoni, E. Specht, M. P. Paranthaman, T. Aytug, A. Goyal, and D. Lee. Enhanced flux pinning by $BaZrO_3$ and $(Gd,Y)_2O_3$ nanostructures in metal organic chemical vapor deposited GdYBCO high temperature superconductor tapes. *Applied Physics Letters*, 94(6):062513, Feb. 2009.

[DGK+15] E. Demenčík, F. Grilli, A. Kario, R. Nast, A. Jung, M. Vojenčiak, J. Scheiter, and W. Goldacker. AC Magnetization Loss and Transverse Resistivity of Striated YBCO Coated Conductors. *IEEE Transactions on Applied Superconductivity*, 25(3):8201405, Jun. 2015.

[DGZG16] E. Demenčík, A. Godfrin, V. M. R. Zermeño, and F. Grilli. Determination of I – V Curves of HTS Tapes From the Frequency-Dependent AC Transport Loss. *IEEE Transactions on Applied Superconductivity*, 26(3):9001004, 2016.

[DUT+07] E. Demenčík, P. Ušák, S. Takacs, I. Vavra, M. Polák, G. A. Levin, and P. N. Barnes. Visualization of coupling current paths in striated YBCO-coated conductors at frequencies up to 400 Hz. *Superconductor Science and Technology*, 20(1):87–91, Jan. 2007.

[DVK+14] E. Demenčík, M. Vojenčiak, A. Kario, R. Nast, A. Jung, W. Goldacker, and F. Grilli. AC Loss and Coupling Currents in YBCO Coated Conductors With Varying Number of Filaments. *IEEE Transactions on Applied Superconductivity*, 24(6):6601008, Dec. 2014.

[Eki06] J. W. Ekin. *Experimental Techniques for Low-Temperature Measurements: Cryostat Design, Material Properties, and Superconductor Critical-Current Testing*. Oxford University Press, 2006.

[FDRW+11] U. Floegel-Delor, T. Riedel, D. Wippich, B. Goebel, R. Rothfeld, P. Schirrmeister, F. N. Werfel, A. Usoskin, and A. Rutt. Reel-to-Reel Copper

Electroplating on Pulse Laser Deposition Coated Conductor. *IEEE Transactions on Applied Superconductivity*, 21(3):2984–2987, Jun. 2011.

[FHM$^+$10] D. M. Feldmann, T. G. Holesinger, B. Maiorov, S. R. Foltyn, J. Y. Coulter, and I. Apodaca. Improved flux pinning in $YBa_2Cu_3O_7$ with nanorods of the double perovskite Ba_2YNbO_6. *Superconductor Science and Technology*, 23(9):095004, Sep. 2010.

[Fuj] Fujikura. http://www.fujikura.com/.

[GBM07] F. Grilli, R. Brambilla, and L. Martini. Modeling High-Temperature Superconducting Tapes by Means of Edge Finite Elements. *IEEE Transactions on Applied Superconductivity*, 17(2):3155–3158, Jun. 2007.

[GFH$^+$07] W. Goldacker, A. Frank, R. Heller, S. I. Schlachter, B. Ringsdorf, K-P. Weiss, C. Schmidt, and S. Schuller. ROEBEL Assembled Coated Conductors (RACC): Preparation, Properties and Progress. *IEEE Transactions on Applied Superconductivity*, 17(2):3398–3401, Jun. 2007.

[GFK$^+$09] W. Goldacker, A. Frank, A. Kudymow, R. Heller, A. Kling, S. Terzieva, and C. Schmidt. Status of high transport current ROEBEL assembled coated conductor cables. *Superconductor Science and Technology*, 22(3):034003, Mar. 2009.

[GGJ$^+$16] R. Gyuráki, A. Godfrin, A. Jung, A. Kario, R. Nast, E. Demenčík, W. Goldacker, and F. Grilli. Interfilament Resistance at 77 K in Striated HTS Coated Conductors. *IEEE Transactions on Applied Superconductivity*, 26(8):6603606, Dec. 2016.

[GGP$^+$14] W. Goldacker, F. Grilli, E. Pardo, A. Kario, S. I. Schlachter, and M. Vojenčiak. Roebel cables from REBCO coated conductors: a one-century-old concept for the superconductivity of the future. *Superconductor Science and Technology*, 27(9):093001, Sep. 2014.

[GK16] F. Grilli and A. Kario. How filaments can reduce AC losses in HTS coated conductors: a review. *Superconductor Science and Technology*, 29(8):083002, Aug. 2016.

[GNK$^+$06] W. Goldacker, R. Nast, G. Kotzyba, S. I. Schlachter, A. Frank, B. Ringsdorf, C. Schmidt, and P. Komarek. High current DyBCO-ROEBEL Assembled Coated Conductor (RACC). *Journal of Physics: Conference Series*, 43:901–904, Jun. 2006.

[GPS+14] F. Grilli, E. Pardo, A. Stenvall, D. N. Nguyen, W. Yuan, and F. Gömöry. Computation of Losses in HTS Under the Action of Varying Magnetic Fields and Currents. *IEEE Transactions on Applied Superconductivity*, 24(1):8200433, 2014.

[GŠVS15] F. Gömöry, J. Šouc, M. Vojenčiak, and M. Solovyov. Round Conductor With Low AC Loss Made From High-Temperature Superconducting Tapes. *IEEE Transactions on Applied Superconductivity*, 25(3):8201004, Jun. 2015.

[GSZV14] F. Grilli, F. Sirois, V. M. R. Zermeño, and M. Vojenčiak. Self-Consistent Modeling of the I_c of HTS Devices: How Accurate do Models Really Need to Be? *IEEE Transactions on Applied Superconductivity*, 24(6):8000508, Dec. 2014.

[GVKZ16] F. Grilli, M. Vojenčiak, A. Kario, and V. M. R. Zermeño. HTS Roebel Cables: Self-Field Critical Current and AC Losses Under Simultaneous Application of Transport Current and Magnetic Field. *IEEE Transactions on Applied Superconductivity*, 26(4):4803005, Jun. 2016.

[HAHN17] S. Hellmann, M. Abplanalp, L. Hofstetter, and M. Noe. Manufacturing of a 1-MVA-class superconducting fault current limiting transformer with recovery-under-load capabilities. *IEEE Transactions on Applied Superconductivity*, 27(4):5500305, Jun. 2017.

[Hal70] M. R. Halse. AC face field losses in a type II superconductor. *Journal of Physics D: Applied Physics*, 3(5):717–720, May 1970.

[HBW+04] T. Haugan, P. N. Barnes, R. Wheeler, F. Meisenkothen, and M. Sumption. Addition of nanoparticle dispersions to enhance flux pinning of the $YBa_2Cu_3O_{7-\chi}$ superconductor. *Nature*, 430(7002):867–870, 2004.

[HDM+09] S. A. Harrington, J. H. Durrell, B. Maiorov, H. Wang, S. C. Wimbush, A. Kursumovic, J. H. Lee, and J. L. MacManus-Driscoll. Self-assembled, rare earth tantalate pyrochlore nanoparticles for superior flux pinning in $YBa_2Cu_3O_{7-\delta}$ films. *Superconductor Science and Technology*, 22(2):022001, Feb. 2009.

[HHJ+04] W. V. Hassenzahl, D.W. Hazelton, B.K. Johnson, P. Komarek, M. Noe, and C.T. Reis. Electric power applications of superconductivity. *IEEE Transactions on Applied Superconductivity*, 92(10):1655–1674, Oct. 2004.

[HKK+19] S. Hahn, K. Kim, K. Kim, X. Hu, T. Painter, I. Dixon, S. Kim, K. R. Bhattarai, S. Noguchi, J. Jaroszynski, and D. C. Larbalestier. 45.5-tesla direct-current magnetic field generated with a high-temperature superconducting magnet. *Nature*, 570(4), Jun. 2019.

[HZKS15] D. Hazelton, Y. Zhang, A. Knoll, and H. Sakamoto. Continuous Improvement at SuperPower of 2G HTS Wire for Demanding Applications, 2015.

[Int06] International Standard. IEC 61788-3 Ed.2: Superconductivity - Part 3: Critical current measurement - DC critical current of Ag- and/or Ag alloy-sheathed Bi-2212 and Bi-2223 oxide superconductors, 2006.

[KGL+06] S. Kang, A. Goyal, J. Li, A. A. Gapud, L. Heatherly, J. R. Thompson, D. K. Christen, F. A. List, M. Paranthaman, and D.F. Lee. High-Performance High-Tc Superconducting Wires. *Science*, 31:1911–1914, 2006.

[KIN+02] T. Kiss, M. Inoue, S. Nishimura, T. Kuga, T. Matsushita, Y. Iijima, K. Kakimoto, T. Saitoh, S. Awaji, K. Watanabe, and Y. Shiohara. Angular dependence of critical current properties in YBCO coated tape under high magnetic field up to 18 T. *Physica C: Superconductivity*, 378-381(PART 2):1113–1117, Oct. 2002.

[KKL+16] I. Kesgin, N. Khatri, Y. Liu, L. Delgado, E. Galstyan, and V. Selvamanickam. Influence of superconductor film composition on adhesion strength of coated conductors. *Superconductor Science and Technology*, 29(1):015003, Jan. 2016.

[KLC+15] I. Kesgin, G. A. Levin, X. Cai, X-F. Li, T. J. Haugan, and V. Selvamanickam. Influence of Oxygenation in Copper Stabilized Multifilamentary 2G HTS Tapes Made by Selective Electroplating and Laser Ablation. *IEEE Transactions on Applied Superconductivity*, 25(3):6601005, Jun. 2015.

[KLHS13] I. Kesgin, G. A. Levin, T. J. Haugan, and V. Selvamanickam. Multifilament, copper-stabilized superconductor tapes with low alternating current loss. *Applied Physics Letters*, 103(25):252603, Dec. 2013.

[KMS13] I. Kesgin, G. Majkic, and V. Selvamanickam. Fully filamentized HTS coated conductor via striation and selective electroplating. *Physica C: Superconductivity*, 486:43–50, Mar. 2013.

[Kru14] P. Kruger. Optimisation of hysteretic losses in high-temperature superconducting wires. *KIT Scientific Publishing, Karlsruhe*, page 218, Feb. 2014.

[KVK+16] A. Kario, M. Vojenčiak, A. Kling, R. Nast, A. Godfrin, F. Grilli, D. C. van der Laan, and W. Goldacker. Comparison of AC losses in Roebel assembled coated conductor (RACC) and CORC® cables assembled using filamentarised conductors at 77 K. *Applied Superconductivity Conference*, 2016.

[LBA+05a] G. A. Levin, P. N. Barnes, N. Amemiya, S. Kasai, K. Yoda, and Z. Jiang. Magnetization losses in multifilament coated superconductors. *Applied Physics Letters*, 86(7):072509, 2005.

[LBA+05b] G. A. Levin, P. N. Barnes, N. Amemiya, S. Kasai, K. Yoda, Z. Jiang, and A. Polyanskii. Magnetization losses in multiply connected $YBa_2Cu_3O_{6+\chi}$-coated conductors. *Journal of Applied Physics*, 98(11):113909, Dec. 2005.

[LBA07] G. A. Levin, P. N. Barnes, and N. Amemiya. Low ac Loss Multifilament Coated Conductors. *IEEE Transactions on Applied Superconductivity*, 17(2):3148–3150, Jun. 2007.

[LBK+06] G. A. Levin, P. N. Barnes, J. W. Kell, N. Amemiya, Z. Jiang, K. Yoda, and F. Kimura. Multifilament $YBa_2Cu_3O_{6+\chi}$-coated conductors with minimized coupling losses. *Applied Physics Letters*, 89(1):012506, Jul. 2006.

[LMH+13] G. A. Levin, J. Murphy, T. J. Haugan, J. Šouc, J. Kováč, and P. Kováč. AC Losses of Copper Stabilized Multifilament YBCO Coated Conductors. *IEEE Transactions on Applied Superconductivity*, 23(3):6600604, Jun. 2013.

[Maw96] Y. Mawatari. Critical state of periodically arranged superconducting-strip lines in perpendicular fields. *Physical Review B*, 54(18):13215–13221, Nov. 1996.

[MCdS+14] W. A. Monteiro, J. A. G. Carrió, C. R. da Silveira, M. Carvalhal, and T. J. Masson. Low Temperature Effect in Electrical Properties of Sintered Copper-Nickel-Aluminum Alloys. *Materials Science Forum*, 805:694–699, Sep. 2014.

[MGC+05] M. Majoros, B. A. Glowacki, A. M. Campbell, G. A. Levin, P. N. Barnes, and M. Polak. AC Losses in Striated YBCO Coated Conductors. *IEEE Transactions on Applied Superconductivity*, 15(2):2819–2822, Jun. 2005.

[MM10] K. Matsumoto and P. Mele. Artificial pinning center technology to enhance vortex pinning in YBCO coated conductors. *Superconductor Science and Technology*, 23(1):014001, Jan. 2010.

[MMB+13] J. P. Murphy, M. J. Mullins, P. N. Barnes, T. J. Haugan, G. A. Levin, M. Majoros, M. D. Sumption, E. W. Collings, M. Polak, and P. Mozola. Experiment Setup for Calorimetric Measurements of Losses in HTS Coils Due to AC Current and External Magnetic Fields. *IEEE Transactions on Applied Superconductivity*, 23(3):4701505, Jun. 2013.

[MMWS13] G. P. Mikitik, Y. Mawatari, A. T. S. Wan, and F. Sirois. Analytical Methods and Formulas for Modeling High Temperature Superconductors. *IEEE Transactions on Applied Superconductivity*, 23(2):8001920, Apr. 2013.

[MSB+07] J. F. Maguire, F. Schmidt, S. Bratt, T. E. Welsh, J. Yuan, A. Allais, and F. Hamber. Development and Demonstration of a HTS Power Cable to Operate in the Long Island Power Authority Transmission Grid. *IEEE Transactions on Applied Superconductivity*, 17(2):2034–2037, Jun. 2007.

[MSCvdL14] M. Majoros, M. D. Sumption, E. W. Collings, and D. C. van der Laan. Magnetization losses in superconducting YBCO conductor-on-round-core (CORC) cables. *Superconductor Science and Technology*, 27(12):125008, 2014.

[Mül97] K.-H. Müller. AC power losses in flexible thick-film superconducting tapes. *Physica C: Superconductivity*, 281(1):1–10, Jul. 1997.

[MZX+09] M. Marchevsky, E. Zhang, Y-Y. Xie, V. Selvamanickam, and G. Pethuraja. AC Losses and Magnetic Coupling in Multifilamentary 2G HTS Conductors and Tape Arrays. *IEEE Transactions on Applied Superconductivity*, 19(3):3094–3097, Jun. 2009.

[NBS+16] R. Nakasaki, P. Brownsey, A. Sundaram, Y. Zhang, D. Hazelton, H. Sakamoto, and T. Fukushima. Progress of 2G HTS Wire Development at SuperPower. In *Applied Superconductivity Conference (ASC 2016)*, Denver, CO, USA, Sep. 2016.

[Nor70] W T Norris. Calculation of hysteresis losses in hard superconductors carrying ac: isolated conductors and edges of thin sheets. *Journal of Physics D: Applied Physics*, 3(4):489–507, Apr. 1970.

[NS07] M. Noe and S. Steurer. High-temperature superconductor fault current limiters: concepts, applications, and development status. *Superconductor Science and Technology*, 20(3):R15–R29, 2007.

[NVD+14] R. Nast, M. Vojenčiak, E. Demenčík, A. Kario, B. Ringsdorf, A. Jung, B. Runtsch, F. Grilli, and W. Goldacker. Influence of laser striations on the properties of coated conductors. *Journal of Physics: Conference Series*, 507(2):022023, May 2014.

[NZB+15] R. Nakasaki, Y. Zhang, P. Brownsey, A. Sundaram, D. Hazelton, H. Sakamoto, and T. Fukushima. Continuous Improvements in Performance and Quality of 2G HTS Wires Produced by IBAD-MOCVD for Coil Applications. In *International Conference on Magnet Technology (MT24)*, Seoul, South Korea, Oct. 2015.

[Ope] The Open University – Superconductivity Course, available at http://www.open.edu/openlearn/science-maths-technology/engineering-and-technology/engineering/superconductivity/content-section-0.

[PAO16] J. C. Prestigiacomo, R. C. Y. Auyeung, and M. S. Osofsky. Modeling the Resistive Critical Current Transitions of Striated Superconducting Tapes. *IEEE Transactions on Applied Superconductivity*, 26(8):6900409, Dec. 2016.

[PKC+02] M. Polák, L. Krempaský, Š. Chromik, D. Wehler, and B. Moenter. Magnetic field in the vicinity of YBCO thin film strip and strip with filamentary structure. *Physica C: Superconductivity*, 372-376(PART 3):1830–1834, Aug. 2002.

[PKM+07] M. Polák, J. Kvitkovic, P. Mozola, P. N. Barnes, and G. A. Levin. Characterization of Individual Filaments in a Multifilamentary YBCO Coated Conductor. *IEEE Transactions on Applied Superconductivity*, 17(2):3163–3166, Jun. 2007.

[PUJ+06] M. Polák, E. Usak, L. Janšák, E. Demenčík, G. A. Levin, P. N. Barnes, D. Wehler, and B. Moenter. Coupling losses and transverse resistivity of multifilament YBCO coated superconductors. *Journal of Physics: Conference Series*, 43:591–594, Jun. 2006.

[PVGŠ11] E. Pardo, M. Vojenčiak, F. Gömöry, and J. Šouc. Low-magnetic-field dependence and anisotropy of the critical current density in coated conductors. *Superconductor Science and Technology*, 24(6):065007, Jun. 2011.

[Rey15] C. Rey. *Superconductors in the Power Grid*. Elsevier Science, 2015.

[RP01] S. Reyntjens and R. Puers. A review of focused ion beam applications in microsystem technology. *Journal of Micromechanics and Microengineering*, 11(4):287–300, Jul. 2001.

[SCB05] M. D. Sumption, E. W. Collings, and P. N. Barnes. AC loss in striped (filamentary) YBCO coated conductors leading to designs for high frequencies and field-sweep amplitudes. *Superconductor Science and Technology*, 18(1):122–134, 2005.

[SCX$^+$09] V. Selvamanickam, Y. Chen, J. Xie, Y. Zhang, A. Guevara, I. Kesgin, G. Majkic, and M. Martchevsky. Influence of Zr and Ce doping on electromagnetic properties of (Gd,Y)–Ba–Cu–O superconducting tapes fabricated by metal organic chemical vapor deposition. *Physica C: Superconductivity*, 469(23-24):2037–2043, Dec. 2009.

[SCZ$^+$12] V. Selvamanickam, Y. Chen, Y. Zhang, A. Guevara, T. Shi, Y. Yao, G. Majkic, C. Lei, E. Galtsyan, and D. J. Miller. Effect of rare-earth composition on microstructure and pinning properties of Zr-doped (Gd,Y)$Ba_2Cu_3O_\chi$ superconducting tapes. *Superconductor Science and Technology*, 25(4):045012, Apr. 2012.

[ŠGK$^+$13] J. Šouc, F. Gömöry, J. Kováč, R. Nast, A. Jung, M. Vojenčiak, F. Grilli, and W. Goldacker. Low AC loss cable produced from transposed striated CC tapes. *Superconductor Science and Technology*, 26(7):075020, 2013.

[ŠGV05] J. Šouc, F. Gömöry, and M. Vojenčiak. Calibration free method for measurement of the AC magnetization loss. *Superconductor Science and Technology*, 18(5):592–595, May 2005.

[SLC$^+$03] M. D. Sumption, E. Lee, C. B. Cobb, P. N. Barnes, T. J. Haugan, J. Tolliver, C. E. Oberly, and E. W. Collings. Hysteretic loss vs. filament width in thin YBCO films near the penetration field. *IEEE Transactions on Applied Superconductivity*, 13(2):3553–3556, Jun. 2003.

[SMM05] R. A. Serway, C. J. Moses, and C. A. Moyer. *Modern Physics*. Brooks Cole – Thomson Learning, 2005.

[SPŠ$^+$13] M. Solovyov, E. Pardo, J. Šouc, F. Gömöry, M. Skarba, P. Konopka, M. Pekarčíková, and J. Janovec. Non-uniformity of coated conductor tapes. *Superconductor Science and Technology*, 26(11):115013, Nov. 2013.

[SŠG14] M. Solovyov, J. Šouc, and F. Gömöry. AC loss properties of single-layer CORC cables. *Journal of Physics: Conference Series*, 507(2):022034, May 2014.

[STI] STI. http://www.suptech.com/.

[SuN] SuNAM. http://www.i-sunam.com/.

[Supa] American Superconductor. http://www.amsc.com/.

[Supb] SuperOx. http://www.superox.ru/en/.

[Supc] SuperPower. http://www.superpower-inc.com/.

[SXL+14] V. Selvamanickam, A. Xu, Y. Liu, N. D. Khatri, C. Lei, Y. Chen, E. Galstyan, and G. Majkic. Correlation between in-field critical currents in Zr-added (Gd,Y)$Ba_2Cu_3O_\chi$ superconducting tapes at 30 and 77 K. *Superconductor Science and Technology*, 27(5):055010, May 2014.

[TC07] O. Tsukamoto and M. Ciszek. AC magnetization losses in striated YBCO-123/Hastelloy coated conductors. *Superconductor Science and Technology*, 20(10):974–979, Oct. 2007.

[Tec] Advanced Conductor Technologies. http://advancedconductor.com/.

[THE] THEVA. http://theva.com.

[The16] The Robinson Research Institute of Victoria University of Wellington. A high-temperature superconducting wire critical current database, 2016.

[TVG+11] S. Terzieva, M. Vojenčiak, F. Grilli, R. Nast, J. Šouc, W. Goldacker, A. Jung, A. Kudymow, and A. Kling. Investigation of the effect of striated strands on the AC losses of 2G Roebel cables. *Superconductor Science and Technology*, 24(4):045001, 2011.

[UKSK09] D. Uglietti, H. Kitaguchi, Seyong C., and T. Kiyoshi. Angular Dependence of Critical Current in Coated Conductors at 4.2 K and Magnet Design. *IEEE Transactions on Applied Superconductivity*, 19(3):2909–2912, Jun. 2009.

[VBW+08] C. V. Varanasi, J. Burke, H. Wang, J. H. Lee, and P. N. Barnes. Thick $YBa_2Cu_3O_{7-\chi}$ + $BaSnO_3$ films with enhanced critical current density at high magnetic fields. *Applied Physics Letters*, 93(9):092501, Sep. 2008.

[vdL09] D. C. van der Laan. $YBa_2Cu_3O_{7-\delta}$ coated conductor cabling for low ac-loss and high-field magnet applications. *Superconductor Science and Technology*, 22(6):065013, Jun. 2009.

[vdLE07] D. C. van der Laan and J. W. Ekin. Large intrinsic effect of axial strain on the critical current of high-temperature superconductors for electric power applications. *Applied Physics Letters*, 90(5):052506, Jan. 2007.

[vdLNM+13] D. C. van der Laan, P. D. Noyes, G. E. Miller, H. W. Weijers, and G. P. Willering. Characterization of a high-temperature superconducting conductor on round core cables in magnetic fields up to 20 T. *Superconductor Science and Technology*, 26(4):045005, Apr. 2013.

[vdLWN+16] D. C. van der Laan, J. D. Weiss, P. Noyes, U. P. Trociewitz, A. Godeke, D. Abraimov, and D. C. Larbalestier. Record current density of 344 A mm^{-2} at 4.2 K and 17 T in CORC® accelerator magnet cables. *Superconductor Science and Technology*, 29(5):055009, May 2016.

[VGT+11] M. Vojenčiak, F. Grilli, S. Terzieva, W. Goldacker, M. Kováčová, and A. Kling. Effect of self-field on the current distribution in Roebel-assembled coated conductor cables. *Superconductor Science and Technology*, 24(9):095002, Sep. 2011.

[VKR+15] M. Vojenčiak, A. Kario, B. Ringsdorf, R. Nast, D. C. van der Laan, J. Scheiter, A. Jung, B. Runtsch, F. Gömöry, and W. Goldacker. Magnetization ac loss reduction in HTS CORC® cables made of striated coated conductors. *Superconductor Science and Technology*, 28(10):104006, Oct. 2015.

[VM95] G. Vellego and P. Metra. An analysis of the transport losses measured on HTSC single-phase conductor prototypes. *Superconductor Science and Technology*, 8(6):476–483, Jun. 1995.

[WGZ+10] S. H. Wee, A. Goyal, Y. L. Zuev, C. Cantoni, V. Selvamanickam, and E. D. Specht. Formation of Self-Assembled, Double-Perovskite, Ba_2YNbO_6 Nanocolumns and Their Contribution to Flux-Pinning and J_c in Nb-Doped $YBa_2Cu_3O_{7-\delta}$ Films. *Applied Physics Express*, 3(2):023101, Feb. 2010.

[WS17] S. C. Wimbush and N. M. Strickland. A Public Database of High-Temperature Superconductor Critical Current Data. *IEEE Transactions on Applied Superconductivity*, 27(4):6900409, Jun. 2017.

[ZCMD94] E. Zeldov, J. R. Clem, M. McElfresh, and M. Darwin. Magnetization and transport currents in thin superconducting films. *Physical Review B*, 49(14):9802–9822, Apr. 1994.

[ZHSG17] V. M. R. Zermeño, K. Habelok, M. Stepien, and F. Grilli. A parameter-free method to extract the superconductor's $J_c(B,\theta)$ field-dependence from in-field current–voltage characteristics of high temperature superconductor tapes. *Superconductor Science and Technology*, 30(3):034001, Mar. 2017.

[ZLF+13] Y. Zhang, T.F. Lehner, T. Fukushima, H. Sakamoto, and D.W. Hazelton. Progress in production and performance of second generation (2G) HTS wire for practical applications. In *IEEE 2013 International Conference on Applied Superconductivity and Electromagnetic Devices*, Beijing, China, Oct. 2013.

[ZSC+09] Y. Zhang, E. D. Specht, C. Cantoni, D. K. Christen, J. R. Thompson, J. W. Sinclair, A. Goyal, Y. L. Zuev, T. Aytug, M. P. Paranthaman, Y. Chen, and V. Selvamanickam. Magnetic field orientation dependence of flux pinning in (Gd,Y)$Ba_2Cu_3O_{7-\chi}$ coated conductor with tilted lattice and nanostructures. *Physica C: Superconductivity*, 469(23-24):2044–2051, Dec. 2009.

Karlsruher Schriftenreihe zur Supraleitung

Karlsruher Institut für Technologie (KIT) | ISSN 1869-1765

Band 001	**Christian Schacherer** Theoretische und experimentelle Untersuchungen zur Entwicklung supraleitender resistiver Strombegrenzer. ISBN 978-3-86644-412-6
Band 002	**Alexander Winkler** Transient behaviour of ITER poloidal field coils. ISBN 978-3-86644-595-6
Band 003	**André Berger** Entwicklung supraleitender, strombegrenzender Transformatoren. ISBN 978-3-86644-637-3
Band 004	**Christoph Kaiser** High quality Nb/Al-AlOx/Nb Josephson junctions. Technological development and macroscopic quantum experiments. ISBN 978-3-86644-651-9
Band 005	**Gerd Hammer** Untersuchung der Eigenschaften von planaren Mikrowellen-resonatoren für Kinetic-Inductance Detektoren bei 4,2 K. ISBN 978-3-86644-715-8
Band 006	**Olaf Mäder** Simulationen und Experimente zum Stabilitätsverhalten von HTSL-Bandleitern. ISBN 978-3-86644-868-1
Band 007	**Christian Barth** High Temperature Superconductor Cable Concepts for Fusion Magnets. ISBN 978-3-7315-0065-0

Die Bände sind unter www.ksp.kit.edu als PDF frei verfügbar oder als Druckausgabe bestellbar.

Band 008	**Axel Stockhausen** Optimization of Hot-Electron Bolometers for THz Radiation. ISBN 978-3-7315-0066-7
Band 009	**Petra Thoma** Ultra-fast $YBa_2Cu_3O_{7-x}$ direct detectors for the THz frequency range. ISBN 978-3-7315-0070-4
Band 010	**Dagmar Henrich** Influence of Material and Geometry on the Performance of Superconducting Nanowire Single-Photon Detectors. ISBN 978-3-7315-0092-6
Band 011	**Alexander Scheuring** Ultrabreitbandige Strahlungseinkopplung in THz-Detektoren. ISBN 978-3-7315-0102-2
Band 012	**Markus Rösch** Development of lumped element kinetic inductance detectors for mm-wave astronomy at the IRAM 30 m telescope. ISBN 978-3-7315-0110-7
Band 013	**Johannes Maximilian Meckbach** Superconducting Multilayer Technology for Josephson Devices. ISBN 978-3-7315-0122-0
Band 014	**Enrico Rizzo** Simulations for the optimization of High Temperature Superconductor current leads for nuclear fusion applications. ISBN 978-3-7315-0132-9
Band 015	**Philipp Krüger** Optimisation of hysteretic losses in high-temperature superconducting wires. ISBN 978-3-7315-0185-5

Band 016 **Matthias Hofherr**
Real-time imaging systems for superconducting nanowire single-photon detector arrays.
ISBN 978-3-7315-0229-6

Band 017 **Oliver Näckel**
Development of an Air Coil Superconducting Fault Current Limiter.
ISBN 978-3-7315-0526-6

Band 018 **Christoph M. Bayer**
Characterization of High Temperature Superconductor Cables for Magnet Toroidal Field Coils of the DEMO Fusion Power Plant.
ISBN 978-3-7315-0605-8

Band 019 **Shengnan Zou**
Magnetization of High Temperature Superconducting Trapped-Field Magnets.
ISBN 978-3-7315-0715-4

Band 020 **Ilya Charaev**
Improving the Spectral Bandwidth of Superconducting Nanowire Single-Photon Detectors (SNSPDs).
ISBN 978-3-7315-0745-1

Band 021 **Juliane Raasch**
Electrical-field sensitive $YBa_2Cu_3O_{7-x}$ detectors for real-time monitoring of picosecond THz pulses.
ISBN 978-3-7315-0786-4

Band 022 **Yingzhen Liu**
Design of a superconducting DC wind generator.
ISBN 978-3-7315-0796-3

Band 023 **Sebastian Hellmann**
Research and Technology Development on Superconducting Current Limiting Transformers.
ISBN 978-3-7315-0804-5

Die Bände sind unter www.ksp.kit.edu als PDF frei verfügbar oder als Druckausgabe bestellbar.

Band 024 **Simon J. Otten**
Characterisation of REBCO Roebel cables.
ISBN 978-3-7315-0904-2

Band 025 **Julia Brandel**
Supraleitende Einzelphotonenzähler: Optimierung der Zeitauflösung und Anwendungsbeispiele aus der Spektroskopie.
ISBN 978-3-7315-0917-2

Band 026 **Dustin Kottonau, Eugen Shabagin, Wescley T. B. de Sousa, Jörn Geisbüsch, Mathias Noe, Hanno Stagge, Simon Fechner, Hannes Woiton, Thomas Küsters**
Bewertung des Einsatzes supraleitender 380-kV-Kabel.
ISBN 978-3-7315-0927-1

Band 027 **Steffen Dörner**
Multifrequenzausleseverfahren von supraleitenden Einzelphotonen-Detektoren.
ISBN 978-3-7315-0961-5

Band 028 **Michael Merker**
Superconducting integrated THz receiver.
ISBN 978-3-7315-0970-7

Band 029 **Wolfgang-Gustav Ekkehart Schmidt**
Superconducting Nanowire Single-Photon Detectors for Quantum Photonic Integrated Circuits on GaAs.
ISBN 978-3-7315-0980-6

Band 030 **Dustin Kottonau, Eugen Shabagin, Wescley de Sousa, Jörn Geisbüsch, Mathias Noe, Hanno Stagge, Simon Fechner, Hannes Woiton, Thomas Küsters**
Evaluation of the Use of Superconducting 380 kV Cable.
ISBN 978-3-7315-1026-0

Band 031 **Alan Preuß**
Development of high-temperature superconductor cables for high direct current applications.
ISBN 978-3-7315-1041-3

Band 032 **Roland Gyuráki**
Fluorescent thermal imaging method for investigating transient effects in high-temperature superconductor tapes and coils.
ISBN 978-3-7315-1064-2

Band 033 **Aurélien Godfrin**
AC Losses in High-Temperature Superconductor Tapes and Cables for Power Applications.
ISBN 978-3-7315-1096-3